CRASHES, CRISES, AND CALAMITIES

Also by LEN FISHER

The Perfect Swarm:
The Science of Complexity in Everyday Life

Rock, Paper, Scissors:
Game Theory in Everyday Life

Weighing the Soul:
Scientific Discovery from the Brilliant to the Bizarre

How to Dunk a Doughnut:
The Science of Everyday Life

CRASHES, CRISES, AND CALAMITIES

HOW WE CAN USE SCIENCE TO READ THE EARLY-WARNING SIGNS

Len Fisher

BASIC BOOKS
A Member of the Perseus Books Group
New York

Published by Basic Books,
A Member of the Perseus Books Group

Books published by Basic Books are available at special discounts for bulk purchases in the United States by corporations, institutions, and other organizations. For more information, please contact the Special Markets Department at the Perseus Books Group, 2300 Chestnut Street, Suite 200, Philadelphia, PA 19103, or call (800) 810-4145, ext. 5000, or e-mail special.markets@perseusbooks.com.

Text set in 10.5 point Weidemann Book

Library of Congress Cataloging-in-Publication Data
Fisher, Len.
Crashes, crises, and calamaties : how we can use science to read the early-warning signs / Len Fisher.
p. cm.
ISBN: 978-0-465-02102-4 (hardback)
1. Accidents—Prevention. 2. Natural disasters—Forecasting. 3. Science—History. 4. Science—Social aspects. 5. Scientists. I. Title.
HV675.F49 2011
363.34'63—dc22

2010048289

E-book ISBN: 978-0-465-02335-6

10 9 8 7 6 5 4 3 2 1

To Wendella, who has survived yet another book without crashes, crises, or calamities.

CONTENTS

ACKNOWLEDGMENTS

My background and contacts in science have given me privileged access to the specialists who are contributing to the rapid development of forecasting methods. Many of them have been kind enough to offer perceptive criticisms, suggestions, corrections, and information in areas where I am not a specialist. Other friends and colleagues have read the manuscript from the nonspecialist's point of view, and their suggestions have contributed considerably to its clarity. I must especially thank my wife, Wendella, who has carefully gone through every page on behalf of the eventual reader, and my agent Barbara Levy for her constant support and belief in what I am trying to do to help integrate science into our wider culture.

Without the help of all these people, this book would not have been possible in its present form. In alphabetical order, those who have made helpful suggestions or provided valuable information are:

Michael Adam, Scott Arthur, Nicola Beech, William Brock, Trish Brown, Steve Carpenter, David Dacam, Underwood Dudley, David Fisher, Wendella Fisher, Eric Foner, Gerd Gigerenzer, Garry Graham, Stephen Guastello, Paul Halpern, Dirk Helbing, Wilhelm Krücken, Alan Lane, Leon Lederman, Matthys Levy, John Lienhard, Rosalinda Madara, Spyros Makridakis, Richard C. Malley, Robert May, Heather Mewton, Marion Mittermaier, Jeff Odell, Roger Pearse, Mark Peterson, Andrew Pyle, Guy Raffa, Mike Retzer, Paul Rosch, Harry Rothman, Marten Scheffer, Alistair Sharp, Ian Stewart, Noel Swerdlow, Graham Turner, Phil Vardy, Charlie Warwick, and Beth Wohlgemuth.

If I have omitted anyone, I hope that they will forgive the unintentional oversight, which I am very happy to correct on my website and in any future editions.

With more than the usual nod toward "any mistakes that remain are my own," I must acknowledge that, despite the help of all these kind people, it is quite possible that, in my efforts to understand, simplify, and make sense of complex material, I may on occasion have been drawn into error. I would be more than pleased to have errors and misunderstandings pointed out in what I hope will be an ongoing dialogue on my website (www.lenfisherscience.com) and elsewhere.

A USER'S GUIDE TO THIS BOOK

This book is designed to be dipped into. While the chapters are presented in chronological order of discovery, they can be read in virtually any order, depending on the reader's interests. My aim is to provide solid information about how we can detect and use early-warning signs to anticipate personal and global disasters. Almost any chapter will have relevant, interesting and often unexpected information for readers to digest and use.

If you want the latest news on global warming, for example, you might like to start with Chapter 9, which tells the story of how models have been developed to help us predict future events in complex personal and global scenarios. If you want an up-to-date list of early-warning signs relevant to our daily lives, you might even turn straight to Chapter 11, which gives such a list, beginning with the little-known fact that the height of women's hemlines is a remarkably accurate warning sign of changing economic circumstances.

The task of unearthing this information, putting it into familiar contexts, and applying it to problems that affect us all has been endlessly fascinating. I hope that you enjoy sharing my journey of discovery.

INTRODUCTION: HOW DO TOADS PREDICT EARTHQUAKES?

The clever men at Oxford
Know all that there is to be knowed.
But they none of them know one half as much
As intelligent Mr. Toad!

—Kenneth Grahame, *The Wind in the Willows* (1908)

On April 1, 2009, the toads in San Ruffino Lake in central Italy left their traditional breeding grounds and headed for the hills. Five days later, a violent earthquake hit the region, demolishing the nearby medieval town of L'Aquila and killing more than three hundred people.

By a singular act of serendipity, the British ecologist Rachel Grant happened to be studying the mating behavior of the toads at the time of the earthquake. She was understandably annoyed when the toads suddenly disappeared from her study site, but thrilled when they returned the day after the earthquake—not just because she could continue her study, but because she was the first scientist to be able to confirm the many anecdotal reports over the centuries of animals acting strangely just before a natural disaster.

Following her scientific account of the phenomenon, there have been suggestions that we could use the behavior of toads or other animals as early-warning signs for earthquakes. Whether this suggestion will withstand scrutiny is not clear, but at least we may be able to discover any physical signals that the animals are responding to and monitor those.

What *is* clear is that the detection of early-warning signals can make a huge difference to our ability to deal with upcoming crashes, crises, and calamities—not only in the natural world but also in our personal lives and in our social and economic environments. This is especially true for the sorts of events that scientists have labeled as *critical transitions.*

In these events, things have been going along smoothly, or changing at a comfortable, steady pace, until abruptly, without apparent warning, there is a jump to a very different state. A volcano explodes; a market collapses; a bridge falls down; a relationship blows up; an epidemic takes off; war breaks out. All of these events, in the scientists' terminology, are critical transitions.

In this book I tell the story of mankind's search for warning signs of such events, and how it has culminated with the recent discovery of a set of early-warning signs that are common to critical transitions of all kinds. I show how these exciting discoveries could be of inestimable value in helping us to predict and deal with sudden crashes, crises, and calamities, both in our personal lives and in the world around us.

Mankind has sought warning signs of such events for millennia. Some cultures put their trust in seers and oracles who were believed to be able to "see" the future. Others looked for omens and harbingers of disaster in the form of unusual heavenly events such as eclipses—a belief that strongly resonates with the modern-day belief in the power of economic and social forecasters, who search for unusual social and economic indicators to guide their prophecies.

With the coming of the scientific era, physical laws began to be uncovered that could act as valuable aids to prediction. Galileo was the first off the blocks when he used simple physical principles to calculate how thick the roof of hell would have to be so that it did not collapse in on itself. Later scientists built on Galileo's physical ideas to develop the concept of *stress,* which engineers use to help predict whether structures like bridges and buildings will be able to support their own weight. This was a considerable advance on the medieval approach of trial and error: a great many cathedrals collapsed during construction or under the influence of the first high wind because the concept of stress was not fully understood.

The physical concept of stress has been "borrowed" by psychologists as an aid to understanding our response to different circumstances. They use it as a measure of the effect that difficult circumstances can have on us, and whether we can cope, or whether we might collapse like the medieval cathedrals.

Stress is not the only conceptual tool that psychologists have borrowed from the physical sciences. An equally important one is *feedback*, which forms the core of our present understanding about how critical transitions arise.

Feedback comes in two forms—positive and negative. Positive feedback reinforces change; negative feedback damps change down and restores equilibrium. Positive feedback can lead to runaway collapse. Negative feedback maintains the balance.*

Positive feedback works by reinforcing change, with the strength of the reinforcement increasing as the change increases. This means that the rate of change continuously accelerates—an apparently insignificant initial change can grow to become a catastrophically big one. A ladder that starts to tilt and fall while you are at the top, growing panic in a crowd, a run on a bank, the growth of an avalanche from tiny beginnings, the evolution of an arms race, the runaway collapse of a fragile ecosystem, the increasing violence of arguments in a deteriorating relationship—these are all examples of positive feedback.

---

* This definition of *feedback* in the physical sense is different from how psychologists define the term; it is the latter usage that has found its way into everyday speech. I clarify the difference in Chapters 4 to 6.

Negative feedback works by providing a restoring force that becomes stronger when change threatens to become larger—like the governor on an engine, which gradually closes off the fuel supply as the engine speeds up, or the way we steer a car by turning the wheel to correct any deviations from the line that we want the car to take. Our bodies also use negative feedback in many ways to maintain *homeostasis*. The hotter we get, for example, the more we perspire, and the greater becomes the cooling effect as the perspiration evaporates.

On a larger scale, it has been proposed that the "balance of nature" and the "invisible hand" of free market competition provide long-term stability by introducing negative feedback into ecosystems and economies, respectively. Unfortunately, it's a myth in both cases. Such negative feedback processes can provide temporary stability, but in the long term our ecosystems and our economies, like our bodies, our relationships, and our societies, are all governed by a complex and constantly shifting balance of positive and negative feedback processes.

Critical transitions happen when positive feedback or some other "runaway" process (such as a buildup of stress) takes over from the normal balancing processes of negative feedback. Our problem in predicting personal, social, economic, and natural disasters is learning how to tell when the balance is stable, on the one hand, and when we are getting dangerously close to a point of instability, on the other. When we get to such a point, intolerable stresses may provoke a sudden collapse, and runaway processes such as positive feedback may suddenly take over and cause the system to run out of control and fail catastrophically.

The underlying processes that produce critical transitions have a lot in common, but until recently the task of predicting the imminence of a critical transition has seemed close to impossible. Sometimes it *is* genuinely impossible, but in the last decade new insights and the development of increasingly powerful computers have allowed scientists to look more deeply into the question. What they have seen offers fresh hope. Out of a dizzying mess of action and reaction, deviation and correction, process and counterprocess, there has come into focus a remarkable set of universal early-warning signs that tell us when situations are about to become critical and when runaway processes are about to take over.

The signs are similar no matter whether we are talking about social disruption, economic disaster, ecosystem collapse, or climate change and other natural catastrophes. Most of the signs are also *simple*—simple enough for all of us to understand and take heed of.

Such early-warning signs promise to let us do Monday morning quarterbacking on a Saturday morning by using discoveries from physics, mathematics, and the world of nature to forecast and handle sudden shocks, surprises, and catastrophes, whether in our personal lives or in the world around us.

When I first came across these findings and recognized their profound importance, I realized that this was something we should all know about. I couldn't wait to find out more and to share my discovery.

This book is the result. It tells the story of mankind's search for predictors of disaster and investigates how we might use the recent and still-evolving discovery of a new set of generic early-warning signs to take firmer control of our own future and the future of the planet.

A NOTE TO READERS

This is the third book in a trilogy in which I investigate how we can use results from the so-called hard sciences to understand and alleviate the personal and social problems that we face in today's complex society.*

* The first two were *Rock, Paper, Scissors: Game Theory in Everyday Life* (New York: Basic Books, 2008) and *The Perfect Swarm: The Science of Complexity in Everyday Life* (New York: Basic Books, 2009).

It is a detective story rather than a sermon—a shared journey of discovery into a new and exciting area of science that is helping us understand more fully the sources of some of our most pressing personal and global problems.

My main aim has been to provide stimulating and thought-provoking ideas from areas that may not be familiar to all readers. I have had to prune and simplify in order to clarify; almost every paragraph could have been expanded into a whole chapter, or even a book. In compensation, there are extensive notes for those who want to pursue particular issues in depth. The notes are an eclectic collection of fascinating anecdotes, references, and additional background information that I came across during my research but couldn't easily work into the main text. They provide an additional dimension and are a detailed resource for those who would like to take up any of the issues I address. Some readers of previous books of mine have even written to say that the set of notes is where they start reading!

Each note is linked to a particular point in the main text, but the notes as a whole are designed to be read independently and also to direct the reader to the most interesting and important original references. I have taken some pains to select the most readable references for the nonspecialist and to add comments when needed. Whether you start with the notes or the main text, I hope that this book stimulates you to think in new and creative ways about your own and the world's future.

Len Fisher
Bradford-on-Avon, United Kingdom,
and Blackheath, Australia

PART 1

A POTTED PRE-HISTORY OF PROGNOSTICATION

1

Do Animals Have Crystal Balls?

For animals, the entire universe has been neatly divided into things to (a) mate with, (b) eat, (c) run away from, and (d) rocks.

—Terry Pratchett, *Equal Rites* (1987)

In 373 BC, the Greek city of Helike, situated on the shore of the Peloponnesus peninsula, was hit by a huge earthquake and then drowned by the ensuing tsunami. Five hundred years later, the ruins could still be seen beneath the clear green waters of the Gulf of Corinth. Roman tourists would sail above them to admire the sunken walls and statuary, and the story of Atlantis may have been based on these ruins.

Then Helike disappeared. The site silted over, nothing could be seen of its buildings and statues, and memory of its whereabouts was lost. The city was not rediscovered until 2001, when it was found buried in an ancient lagoon. It is now a marine archaeological site—one of the most endangered in the world.

Memory of the city's location may have been lost, but the circumstances of its passing were recorded by several historians, among them the Roman author Claudius Aelinius (Aelian). His account is particularly important because it contains the first description of animals "predicting" an earthquake. According to Aelian:

> For five days before Helike disappeared, all the mice and martens and snakes and centipedes and beetles and every other creature of that kind in the city left in a body by the road that leads to Keryneia [Corinth].

> And the people of Helike seeing this happening were filled with amazement, but were unable to guess the reason. But after these creatures had departed, an earthquake occurred in the night; the city subsided; an immense wave flooded and Helike disappeared.

There have been many other reports of animals behaving oddly before a tsunami. There are apparently reliable anecdotal reports, for example, of animals fleeing for safety before the disastrous Asian tsunami of December 26, 2004, including stories told to me personally when I was in the area a fortnight later. But what could produce such behavior? Some people believe that animals have a "sixth sense" that enables them to detect forthcoming danger. If this sensitivity to future events really exists, and if it can be harnessed, then forecasting the onset of disasters would suddenly become a whole lot easier.

One example of this sense is the claimed ability of some dogs to "know" when their owners are coming home. One such claim was "investigated" by an Austrian television company, whose film appeared to contain convincing proof that a dog called Jaytee could tell when its owner was setting off to return home from a distant location and would promptly go to the porch and wait. Parallel filming of the dog and the owner appeared to confirm this claim in dramatic fashion.

When the psychologist Richard Wiseman and his colleagues investigated the claim, they found that there could have been many other explanations. In the Austrian program, for example, the person filming the dog's behavior appears to have known of the planned return time and may have given inadvertent signals to the dog. To eliminate this and other possibilities, Wiseman and his colleagues set up a protocol of having the return time selected randomly by an experimenter who accompanied the owner *after* they had left the house, and without informing an observer who had remained in the house to watch the dog's behavior. With this protocol in place, there was no evidence at all for the original claim.

There were even more flaws in the "experiments" involving Paul the octopus during the 2010 World Cup of soccer. Housed in a German aquarium, the animal correctly predicted the result of every match involving the German team (usually a win), and also the result of the final, simply by choosing food from one of two boxes labeled with the appropriate national flags.

The octopus's performance provoked public enthusiasm and criticism in equal measure, especially among the German public—enthusiasm when the octopus predicted a German win (Germany was usually by far the favored team) and outrage when it successfully predicted Germany's ultimate loss to Spain, with some fans calling for the octopus to be cooked and eaten—a classic case of shooting the messenger. This response was nothing, though, compared to that of Iranian president Mahmoud Ahmadinejad, who accused the West of using the octopus to spread "Western propaganda and superstition."

But how did the octopus do it? Before we can possibly countenance claims of psychic powers, we need to follow Wiseman's example in the case of the "psychic dog"—that is, we need to be quite sure that the powers are real and that all reasonable alternative explanations have been eliminated.

In Paul's case, even the powers are in dispute, because he did not do nearly so well in predicting the results of the earlier European Cup (getting only four results correct out of six). There are also plenty of alternative explanations for his later astonishing success. The boxes were transparent, so he could see the flags, and he may have been more attracted to the German flag. There were no precautions in place to ensure that the keeper wasn't giving clues about his own expectations. None of the results was really unexpected, and Germany and Spain (to whom Germany eventually lost) were the pre-tournament favorites. The piece of octopus food (a mussel) may even have been made larger (not necessarily deliberately) in the box that the keeper favored.

All in all, it was a bit of fun, but as a scientific experiment—forget it. In fact, forget all of the so-called proofs that animals have psychic powers. It is incredibly difficult to design a scientific experiment that eliminates all other reasonable explanations and leaves psychic powers as the only alternative.

That's not to say that some animals may not have other special powers. The toads of San Ruffino Lake, for example, may have been sensitive to the increase in very low-frequency radio emissions from the ionosphere that preceded the earthquake; indeed, some seismologists believe that such emissions could act as early-warning signals for many earthquakes. The same seismologists have pointed out, however, that if we could establish a genuine correlation between animal behavior and some physical change, that would mean that we were already able to measure the physical change and thus would have no need to rely on the animal behavior to make our predictions.

COULD HUMAN "PRECOGNITION" HELP US PREDICT THE FUTURE?

With humans, it's different. We can test the possibility of precognition more thoroughly because our ability to verbalize allows us to make more detailed predictions than if we were limited to purely behavioral clues.

One of the first people to attempt such a test was the fabulously wealthy King Croesus, who ruled the ancient kingdom of Lydia (in what is now northern Turkey) from 560 to 546 BC. Croesus wanted to know just how reliable were the predictions of the many "Oracles" then plying their fortune-telling trade, and he decided to investigate their abilities before paying hard cash for their advice. To perform his test, he sent messengers with instructions to ask each Oracle on a prearranged day "what Croesus, son of Alyattes and king of Lydia, was doing at the moment."

What he was in fact doing was boiling up a tortoise and a lamb together in a bronze cauldron with a bronze lid. Unsurprisingly, none of the Oracles came even close to the right answer, with one exception—the famous Oracle who resided at the Temple of Apollo (the remains of which still exist) in the Greek town of Delphi.

The position of Oracle was usually occupied by a middle-aged countrywoman who sat precariously balanced on a three-legged stool above a crack in the ground while she received questions. She would then sniff the volcanic vapors emerging from the crack, go into a trancelike state, and produce a series of semi-comprehensible ravings that were "translated" into elegant Greek hexameters by the attendant priests of Apollo.

Needless to say, there was plenty of opportunity for chicanery, and chicanery is surely the explanation for the answer she sent back to Croesus:

The smell has come to my sense of a hard-skulled tortoise
Boiling and bubbling with lamb's flesh in a bronze pot:
The cauldron underneath is of bronze, and of bronze the lid.

Presumably the messenger had somehow discovered in advance what Croesus would be doing and had dropped a hint to the priests of Apollo in return for a bribe. Whatever the explanation, Croesus was taken right in, and he sent his messenger back with a second question—should he make war on the Persian Empire?

This time the messenger was accompanied by a huge pile of treasure, including a statue of a lion made from pure gold and weighing some 750 pounds. Whatever the original bribe was, it had certainly proved to be a shrewd investment, but the message that Croesus received back was considerably less specific than the first one. It simply said that, if Croesus attacked Persia, "a mighty empire would fall." Croesus took this as a signal to go ahead. Unfortunately, the empire in question turned out to be his.

That was one of the problems with the ancient Oracle's predictions. They were generally ambiguous and usually too vague to be genuinely useful. The same applies to the predictions of her modern successors who ply their trade among the credulous. I tested one modern Internet Oracle whose picture on her home page suggests that she is both younger and more attractive than her ancient counterpart. Sadly, though, the poetry was missing from her predictions. When I asked her, "How well will my next book sell?" the answer was, "Not too well. Please keep your hands on the keyboard at all times during the session." Try translating that into Greek hexameters.

SCIENTIFIC TESTS OF PRECOGNITION

Large rewards are still on offer for anyone who can demonstrate genuine psychic powers. The magician James Randi, for example, has a standing offer of $1 million for a successful demonstration under scientifically controlled conditions. The conditions are no more onerous than any scientist would use if he or she hoped to convince other scientists about the validity of an experiment, and competitors for the prize are invited to participate in the experimental design. The reward has been on offer for forty-six years, but so far, says Randi, none of the more than one thousand people who have challenged for the prize has even got to first base.

A surprising number of people nevertheless believe that there might be "something in" the paranormal. According to the Australian sheep-goat scale, designed to test the extent of our belief in the paranormal, some 30 percent of us even believe that we share the Oracle's supposed ability to see at least dimly into the future. Amazingly, the research of the U.S. cognitive scientist Mark Changizi suggests that the 30 percent (who inhabit the "sheep" end of the scale) may have a case. The bad news is that the furthest that they or the rest of us can "see" is about one-tenth of a second.

Changizi's research concerns what happens when light hits the retina. It takes the brain around one-tenth of a second to translate this information into a visual image. According to Changizi, the brain adapts to this delay by generating "pre-images" of what it believes will occur one-tenth of a second into the future. This foresight keeps our view of the world in the present. It gives us enough time, for example, to catch a fly ball instead of getting socked in the face, or to maneuver our way through a crowd without bumping into people.

Of course, this is not true precognition, even over such a short time scale. The brain *imagines* what is going to happen. It does not *know*. When it comes to knowing the future, we skeptical goats follow Carl Sagan's dictum that "extraordinary claims require extraordinary proof." Precognition is an extraordinary claim, but so far no one has produced the correspondingly extraordinary proof.

There have, however, been claims that initially convinced skeptics. Some of the most convincing experiments were those of the British mathematician Samuel Soal in the early 1940s. A person on one side of an opaque screen was asked to look at a series of cards, while a person on the other side was asked to guess what card the subject was looking at. These experiments did not produce a statistically significant positive result until Soal thought to check out the correlation between the guesses and the card *after* the one that the first subject had looked at. For several guessers he found a very strong correlation, which seemed to show that the guesser had foreknowledge of what card was coming next!

These data were sufficient to convince the Cambridge philosopher C. D. Broad, who wrote an extended review of the experiments for the respected journal *Philosophy* under the title "The Experimental Establishment of Telepathic Precognition." The experiments were especially convincing because Soal had previously used statistical analysis to analyze over 120,000 trials of card-guessing with 160 participants without ever being able to report a significant finding. He had also debunked telepathy in a review where he opined that it was "a merely American phenomenon." Now, it appeared, it was a British one as well.

It was one, however, that he was never able to repeat. Later examination of his data also revealed that some of the most significant results were almost certainly fabricated, to the extent that even committed parapsychologists no longer believe in them.

Why Soal should have fabricated his results is an open guess. Unfortunately, by the time the fabrication was demonstrated, he was suffering from dementia and in no position to defend himself or explain his actions. In any case, there is *no* compelling scientific evidence from Soal's experiments or elsewhere for the existence of precognition, whether by telepathy, dreams, clairvoyance, or any other means. Most reported experiences can be explained in one of five ways, four of which were identified by Broad as long ago as 1937:

1. *Selection bias:* Remembering when events that we have "foreseen" came to pass and forgetting all the times when they did not eventuate.
2. *Cryptomnesia:* Having a forgotten memory return without recognizing it as such and instead believing it to be something new and original. The pioneering psychiatrist Carl Jung argued that this was not only a normal mental process but a necessary one as well. Without this process, he maintained, the human mind would always be cluttered or overloaded with random information.
3. *Unconscious perception:* Unconsciously inferring, from information that has been unconsciously learned, that a certain event will probably happen in a certain context.
4. *Self-fulfilling prophecy:* A situation where an event comes to pass *because* it was "foretold." A child who is told that he

or she is bound to be a failure, for example, may really become a failure just because of this "prophecy."

The fifth explanation (putting aside the issue of straight-out fraud) is based in the regrettably common misuse or misunderstanding of statistics. It is not to deny the experience of those who have had "premonitions" to point out that, statistically speaking, most of these are likely to be no more than coincidence. Robert Todd Carroll, author of *The Skeptic's Dictionary*, explains premonitions in terms of "the Law of Truly Large Numbers":

> Say the odds are a million to one that when a person has a dream of an airplane crash, there is an airplane crash the next day. With 6 billion people having an average of 250 dream themes each per night, there should be about 1.5 million people a day who have dreams that seem clairvoyant.

True believers, though, will always hang on to their belief that premonitions must have "something in them," even though Aristotle dismissed precognition, Freud himself ridiculed the idea that dreams could reveal the future, and modern "clairvoyants" have repeatedly failed scientific tests of their supposed abilities and often been exposed as frauds. There is, in fact, *no* scientific evidence that we can directly visualize the future, but there are plenty of arguments from science and philosophy that it is impossible to do so while still being firmly anchored in the present.

THE COLLIDER, THE PARTICLE, AND A THEORY ABOUT FATE

One of the most powerful arguments from philosophy was advanced by the Roman statesman and philosopher Marcus Tullius Cicero, who poked fun at seers, soothsayers, and fortune-tellers in his remarkable treatise *De Divinatione.* Published in 44 BC, it is one of our most important sources of information on the methods of divination that were popular at the time. It is also an exercise in debunking that the professional debunker James Randi himself would have been proud of.

Cicero's argument is that divining the future is logically impossible, since we would surely act on the information and any such action would immediately change the future. If Cicero himself had divined the possibility of his own assassination, for example, he would undoubtedly have avoided being in the place where it eventually happened.

If Cicero were alive today, he would be pleased to know that his skepticism has received the support of a tongue-in-cheek scientific theory proposed in 2006 by the string theorist Holger Nielsen from the prestigious Niels Bohr Institute in Copenhagen and the theoretical physicist Masao Ninomiya from the Okayama Institute for Quantum Physics in Japan. These authors suggest that it is simply not possible to "see" the future from the perspective of the present, because the future itself prevents us from doing so.

The theory was brought to public attention by the *New York Times* on October 12, 2009, when its readers were treated to a portentous headline: "The Collider, the Particle, and a Theory About Fate." The "particle" was the Higgs boson, a tiny nuclear particle that has never yet been seen, but that is thought by physicists to be responsible for conveying the property of mass to all other particles in nature. The "collider" is a machine built to search for the particle—the Large Hadron Collider (LHC). Housed in a seventeen-mile-long, doughnut-shaped tunnel buried deep beneath the border between France and Switzerland, the LHC is designed to send nuclear particles crashing into each other at near-light-speeds in the hope of releasing the boson.

The "theory about fate" postulated an extraordinary connection between the machine and the boson that it was designed to search for.

The theory advanced the possibility that the Higgs boson can reach back from the future and hide itself from sight by wrecking the experiments that are aimed to discover it.

Don't laugh. Well, okay, do laugh, but not too loud. Nielsen and Ninomiya themselves admit that their speculative model "begins with a series of not completely convincing, but still suggestive assumptions." Given those assumptions, though, the model uses well-established and well-understood physical laws to reach its remarkable conclusions. These conclusions are far from proven, but neither have they been shown to be scientifically impossible.

The fate of the Texas-based Superconducting Super Collider (SSC) bears witness to the possibility that the theory could be right. Designed to search for the Higgs boson, the SSC was started in 1991, but the project was canceled by Congress in 1993 before construction could be completed. Bad luck? Or was it, as the authors propose, "backward causation," with the particle itself acting to stop the project?

The history of the LHC provides further evidence. The first test was performed on September 10, 2008, and my colleagues from Bristol University who had been involved for many years in the design and building of the detectors used by the collider were thrilled with the result. They weren't so thrilled when, nine days later, two of the electromagnets that pushed the particles around failed and the experiments had to be stopped. Bad luck? Or an ominous message from the future?

It took over a year to repair the damage and get the machine up and running again. Recently I received a triumphant e-mail from the Bristol high-energy physicist Dave Newbold, dated March 30, 2010, and entitled "Watch rooms full of excited particle physicists live!" I logged on to the Web address for the European Organization for Nuclear Research (CERN) and was rewarded by seeing Dave on the CERN webcam, waving and smiling as the record of successful high-energy collisions grew steadily. Behind him, though, I could have sworn that I saw the malevolent face of a phantom Higgs boson, waiting to pull the rug out from under the experiments yet again. Time will tell.

There have been some claims that the esoteric reaches of quantum mechanics may make some form of precognition possible, as well as counterclaims that such effects are unprovable, although ideas such as those involving the existence of parallel universes may allow for the possibility. For the moment, though, such ideas belong in the realm of speculative theoretical physics—or perhaps science fiction.

In the real world, true precognition has never been demonstrated in man or other animals, and there are many arguments against the very possibility. We must look elsewhere if we are to gain forewarning of future events. The first place that mankind went looking for such knowledge was in the heavens.

2

The Future Eclipsed

The clouds I can handle, but I can't fight with an eclipse.

—Stephenie Meyer, *Eclipse* (2007)

THE FUTURE IN THE STARS

On October 22, 2134 BC, two Chinese court astronomers called Hsi and Ho lost their heads over an eclipse. They were not the first to do so, nor would they be the last. People have been losing their heads over eclipses since time immemorial, fearing them to be portents of things to come. Usually pretty bad things.

This particular eclipse, one of the first to be the subject of written record, was certainly bad for Hsi and Ho. Their job was to forecast eclipses and other heavenly events. According to some reports, they even had a primitive planetarium to aid them. The dome of the sky was divided into degrees, with the stars represented by pearls and the earth, sun, moon, and planets by precious stones, all being moved by hand to follow their changing positions in the sky.

Unfortunately for Hsi and Ho, a planetarium was not the only thing that they had access to. It appears that they also had an entrée to the emperor's store of wine, which they were busy drinking when the eclipse made its unpredicted appearance. According to the ancient Chinese document *Shu Ching*, "Hsi and Ho, drunk with wine, had made no use of their talents. Without regard to the obligations which they owed the Prince, they abandoned the duties of their office, and they are the first who have troubled the good order of the calendar whose care has been entrusted to them."

One has to have some sympathy. The prediction of eclipses requires more precise observation than was possible at the time, and they were also unlucky in that any particular point on the earth's surface is only likely to experience a full solar eclipse once every four hundred years or so. Surely this leaves time for the odd glass of wine.

Their drunken behavior, however, left the emperor Chung K'ang with an immediate problem—how to deal with the eclipse? Eclipses were thought to be caused by a dragon trying to eat the sun, and the established procedure was to assemble soldiers, courtiers, and whoever else happened to be around to fire arrows, beat drums, and bang pots and pans to frighten the dragon away. Time was short, but we can imagine that Chung K'ang organized what he could along these lines. We can also imagine that he must have felt mightily relieved when the sun reappeared. We don't know for sure. What we do know is that, when it was all over, he summoned Hsi and Ho, who "by their negligence in calculating and in observing the movement of the stars . . . had violated the law of death," and had them beheaded.

This fascinating story is no less relevant for probably having been fabricated around AD 300. Its real point is that ancient civilizations took eclipses very seriously as harbingers of important events. The Chinese saw them as interrupters of balance and regularity and as events that particularly affected the fate of kings. Many other ancient cultures had similar beliefs.

According to the astrophysicist David Dearborn, it made sense for these cultures to rationalize solar eclipses as bad omens. "For most early cultures," he argues, "the sun was seen as a life-giver, something that was there every day, so something that blots out the sun was a terribly bad event, filled with foreboding."

We may no longer feel that sense of foreboding, but eclipses still exert their fascination, as I discovered when I was guiding a tour group around the Large Hadron Collider (LHC) in Geneva and saw a headless astronomer standing outside the building in which it was being constructed.

I thought for a moment that the ghost of Hsi or Ho must have returned. Then I realized that it was one of the senior scientists whom I knew well, but whose head was concealed behind a black welding mask through which he was staring at the sun. "There's a partial eclipse," he said. "Here, have a look."

The members of the group spent as much time staring at that eclipse through a welding mask* as they did in touring the LHC facility. It was an excellent opportunity for me to explain that both the eclipse and the LHC involved events that had already happened. The moon blocks the light from the sun just over a second before we see it. The LHC is concerned with rather older events—those that happened nearly 14 billion years ago, when the universe was just being born and when the Higgs boson is thought to have first put in an appearance.

The LHC experiments are designed to free the hypothetical Higgs boson from nuclear captivity by smashing protons into each other with sufficient violence. According to current theory, we will never be able to see the particle itself, because the freed particle will break down almost instantaneously to produce a shower of other particles. The correlations between the speeds and directions of these particles in the strong magnetic field of the detector have been predicted from theory. They will provide a test of the theory and a "signature" for the ghostly presence of the Higgs boson—if it exists.

---

* Don't try this without specialist guidance!

CONNECTING THE PAST WITH THE FUTURE: CURIOUS CORRELATIONS AND FALLACIOUS FORECASTING

The correlations that the LHC experiments seek will be used to *test* a prediction. The ancient civilizations of Mesopotamia (the Babylonians, Assyrians, and Sumerians) used correlations in a different way—to *make* predictions rather than to test them. The correlations they used were recorded in a set of books now known as the "omen series." Each page contained two columns—one recording unusual heavenly events like eclipses, the other a list of the earthly events with which they were thought to be correlated.

When an eclipse, an occultation, or an unusual arrangement of stars occurred, these ancient peoples would look through their lists to see whether the event might signify the occurrence of a good crop, the death of a king, or even the collapse of an empire. In doing so, they were laying the foundations for the pseudo-science of astrology. They were also laying the foundation for methods that are used by today's highly paid prophets of society and the marketplace (which some commentators liken to astrology).

These methods, like those of the ancients, consist in looking for correlations between different events so that one can be used as a predictor for the other. Today's market gurus, for example, might use a spike in the gold price as an indicator for a coming downturn in the market because these two events have been correlated in the past. Unfortunately, this "correlation" method of forecasting is susceptible to a series of logical fallacies that make it suspect, to say the least, and worse than useless when it comes to forecasting critical transitions.

All of these fallacies are variants of "the Apparent Pattern"—the belief that, if we can perceive a pattern in things, then the pattern *must* have some meaning.

It's a belief that has served us well throughout evolution—in fact, we could hardly have survived without it. The ability to pick patterns out of a mass of information is essential for recognizing faces, learning language, and remembering the geography of our surroundings. But it is also possible to make mistakes. The question is, what sort of mistake should we make? Is it better to believe falsehoods or reject truths?

Our best evolutionary strategy was to accept a few falsehoods so as not to miss an essential truth. A rustle in the grass could have been due to the wind, but it might also have indicated the presence of a menacing carnivore. Those who survived to pass their genes on were the cautious ones who bet on the carnivore, even though most of the time it was the wind.*

The downside of this genetic inheritance is that we have evolved a tendency to believe in the reality of the patterns that we perceive, whether real or imaginary. Unfortunately, we have not yet evolved a built-in "Baloney Detection Network" to help us distinguish between the two. The best that we can do is to check our beliefs out against the three basic philosophical fallacies that are subsumed under the heading of the Apparent Pattern:

Fallacy 1: Post Hoc Ergo Propter Hoc (After This, Therefore Because of This)

"Post hoc ergo propter hoc" is the belief that if A occurs before B, then A *must* be the cause of B. Called "the post hoc fallacy" for short, it was first described by the Greek philosopher Aristotle, a student of Plato and teacher of Alexander the Great, around 350 BC.

---

* These characteristic of the human race were brought into sharp focus by Bobby Henderson's wonderful spoof *Gospel of the Flying Spaghetti Monster* (London: Harper Collins, 2006), which ascribes everything that happens in the universe to the actions of the monster's "noodly appendages." Worshipers of the spaghetti monster are appropriately called *pastafarians*.

My daughter, who is a fellow scientist, produced a marvelous example of this fallacy when she pointed out that farmers were migrating from the Australian state of New South Wales to the state of Victoria and turning into letterboxes. Her "evidence" for this intentionally preposterous proposition was that the number of farmers who left New South Wales over a particular period was precisely equal to the subsequent increase in the number of letterboxes in Victoria during the same period.

When analysts look for "indicators," they are very prone to fall for the post hoc fallacy. There is a strong tendency to believe, for example, that if volatility in the housing market is followed by a drop in share prices, then the volatility must somehow have caused the drop, or else the two are linked by a common factor, which means that the first can be used as a predictor of the second. We can blame such beliefs on evolution, but the bottom line is that, unless there is a real underlying mechanism, such correlations are not to be relied upon. The safest correlation we can use is probably that suggested by the Seattle financial planner Elaine Scoggins, who claims that the surest indicator of a coming stock market collapse is an increasing tendency to believe in the advice of experts.

Fallacy 2: Cherry-Picking

Cherry-picking consists of selecting data to fit your hypothesis or belief system. A necessary strategy for our individual development, it starts when we are infants trying to make sense of the peculiar noises emanating from our parents' mouths. Eventually we sort out the words from the background noise by selecting some patterns and dismissing others. We also learn to cherry-pick with our visual system, to the extent that if we are later faced with an unfamiliar pattern, it is difficult for us to associate it with a real object.

Cherry-picking becomes a problem when we start using it to make predictions and choose the wrong set of cherries. Often we don't even realize that we are doing it. One classic case was the famous *Literary Digest* poll of 1936, which predicted that Republican Alf Landon would beat the incumbent Franklin D. Roosevelt in the presidential election. (Roosevelt actually won by a landslide.) The mistake that the *Digest* made was to conduct its poll by telephone, thus "cherry-picking" voters who could afford to own telephones.

Especially prone to this fallacy are modern-day data-miners, who are searching for "indicators" that may be used to predict future trends or for patterns that may be as innocuous as those that reveal consumer preference or as frightening as those that indicate credit card theft, identity fraud, or even terrorist activity. Unfortunately, the correlations that such searches throw up can often be spurious and misleading.

One of my favorites was a report in the *New York Times* that countries where people spend less time eating have higher economic growth rates. (If you think that this really could be true, think of the amount of time that people in Third World countries are likely to spend at the dinner table.) The "evidence" for this preposterous proposition came from an Organization for Economic Cooperation and Development (OECD) survey of seventeen countries. Two "cherry-picked" groupings were found to have significantly different eating habits. People in Mexico, Canada, and the United States spent an average of 75 minutes per day consuming their grub; residents of New Zealand, France, and Japan spent more than 110 minutes per day on the same activity. The first group had a significantly higher economic growth rate than the second; ergo, according to the newspaper story, the sure way to economic growth is to reduce the time you spend at the dinner table.

This is not to demean the work of professional data-miners, who are generally well aware of these problems and take considerable care to avoid them. But the temptation to cherry-pick is still there, and sometimes one just wonders. . . .

A problem related to cherry-picking is that of the "false positive," which is becoming an increasingly serious issue as the use of data-mining methods to identify terrorist or other criminal activity becomes more widespread. There are substantial issues of personal liberty at stake. Just how many false positives should we accept in order to reduce the chances of mistakenly excluding a "true positive"?

Suppose we have a test that will correctly identify a suspect as guilty eight times out of ten but miss two times out of ten, and that will correctly identify a suspect as innocent nine times out of ten but identify the person as guilty one time in ten. Sounds pretty good? It's certainly better than the performance of so-called lie detector tests, according to a thorough examination of such tests by the U.S. National Academy of Sciences.

When we apply it, though, the results don't make pleasant reading. Say that we *know* our suspect is someone within a group of one hundred people (an industrial spy within a particular arm of a company, for example), and we apply the test to all of them. There is an 80 percent chance that our suspect will be identified as guilty, but there is still not much chance of identifying who that person is, because we will also have ten false positives to contend with. If the culprit is hidden within a group of one thousand, there will on average be one hundred false positives. When the number of false positives begins to dwarf the number of true positives, cherry-picking via data-mining becomes worse than useless—it becomes positively dangerous.

Fallacy 3: The Future Will Be the Same as the Past

When it comes to predicting critical transitions, the most dangerous fallacy of all is the assumption that the future is going to be the same as the past. This assumption works okay when it comes to scientific laws, where one of the primary tests is *reproducibility*. When it comes to predicting the course of events in complex societies, economies, and ecosystems, though, it's hazardous to assume that the future will be the same as the past because, as the decision theorists Spyros Makridakis and Nassim Taleb point out, the future is not like it used to be.

Even for the relatively simple problem of predicting trends they say, "History never repeats itself in exactly the same way. This means that statistical models that extrapolate (or interpolate) past patterns/relationships cannot provide accurate predictions, since they assume that such patterns/relationships will not change."

Even if history did repeat itself exactly, we would still have some difficulty in predicting trends because our knowledge of history is inexact and so our measurement of previous trends cannot be precise. Whether we are talking about trends in the marketplace, in society, or in nature, they are usually *averages*, taken through a scatter of points. When we try to extrapolate these trends too far ahead, we can end up in trouble.

Arguing that the evolution of society has mostly proceeded through events that *nothing* in our past could have prepared us for, Taleb cites examples such as the rise of the Internet, the development of the personal computer, surprise terrorist attacks, the discovery of penicillin, and the many original ideas that seem to arise from nowhere in the arts and the sciences. He labels such rare important events as "Black Swans"* and argues that we cannot extrapolate from past experience to predict them. The best we can do is to create suitable conditions for them to happen (or not to happen, as the case may be).

Taleb has a particularly strong point when it comes to science. There is much evidence that most of the important applications of science have come from Black Swan discoveries and thus have been virtually impossible to predict. To cite just one example, X-rays were discovered

---

* The term "Black Swans" comes from a classical philosophical example that has its origins in seventeenth-century Europe, where the only swans anyone had ever seen were white. "All swans are white" seemed almost self-evident until the Dutch explorer Willem de Vlamingh reached the coast of Western Australia on January 7, 1697, and became the first European to see a *black* swan.

serendipitously by William Röntgen when he was trying to work out how gases could conduct electricity. Not only did his research have nothing to do with the subsequent important application of X-rays in medical diagnosis, but the importance of his discovery (or even its very existence) could not have been predicted when he began his research. The obvious conclusion from this and many similar examples is that, following Taleb, our best strategy is to allow a sufficient number of scientists to ask questions that *they* perceive as important, regardless of whether an immediate application seems likely.

Black Swans are rare important events that involve abrupt, dramatic, and unexpected change. Some, like the discovery of X-rays, are inherently unpredictable. The prior knowledge just isn't there. Others, though, are caused by critical transitions, where the seeds of sudden change are built into a system and a complicated balance between positive and negative feedback processes slowly takes the system to a tipping point where a further small change initiates an abrupt and dramatic shift. As many of the examples in this book will show, the point at which this happens is often predictable in principle, and sometimes in practice, although we are still at an early (and exciting) stage when it comes to understanding these processes in real life.

Engineers, for example, can calculate and measure the stresses in a structure, and they know when those stresses have reached a critical point where the structure is liable to collapse. Ecologists are now learning how to use knowledge of the various positive and negative feedback processes in ecosystems to predict when the slow evolution of the system will take it to a tipping point, such as when a shallow lake "flips" from a clear to a turbid state, or when animal populations alternate between "boom" and "bust." Economists are now learning from ecologists and may one day be able to use similar methods to anticipate sudden market changes.

The learning process goes right back to Galileo, and his efforts to calculate the dimensions of the roof of Hell.

3

Galileo's Hell

Abandon hope, all ye who enter here.

—Dante Alighieri, "Inferno"

Heaven for climate,
and Hell for society.

—Mark Twain, describing a clergyman's response to the question of whether it would be better to end up in Heaven or Hell after dying

Dante's "Inferno" is a fourteenth-century account of a visit to Hell that the sixteenth-century Church of Galileo's time took literally. So literally, in fact, that one of its cardinals asked the twenty-four-year-old Galileo to calculate the exact size of Hell and its denizens from Dante's description. Galileo got the answer embarrassingly wrong, but he kept his mistake secret. When he finally published the right answer near the end of his life, it provided the first step in the true scientific prediction of the critical transitions that can lead to disaster.

History does not record what Galileo thought privately about the cardinal's amazing request. In public, he was all for it. He had recently graduated in mathematics from the University of Pisa, but he was still without a permanent job. This seemed to be an excellent opportunity to impress the influential people who controlled those jobs, and he set to work with a will.

According to Dante, Hell was situated beneath the earth's crust and divided into nine concentric levels. Sinners of increasing wickedness were confined for eternity at progressively deeper levels.*

Galileo concluded from Dante's description that Hell must be shaped like an ice cream cone, with the point at the center of the earth and the vaulted roof (the surface of the ice cream) forming a part of the earth's surface centered on Jerusalem and spanning a diameter equal to the radius of the earth, which he took to be 3,245 miles.

That's a huge vaulted roof. But how thick should it be to support its own weight without collapsing? This was where Galileo made his mistake.

A QUESTION OF SCALE

Galileo presented his results in two public lectures at the prestigious Florentine Academy. He played to his audience by constantly referring to two previous efforts to calculate the dimensions of Hell—one by a Florentine called Antonio Manetti and the other by a non-Florentine called Alessandro Vellutello. He praised Manetti's efforts and waxed more and more satirical about those of Vellutello as his lectures went on. This was just what his Florentine audience wanted to hear, and his performance demonstrates that Galileo was quite an astute politician, at least in the academic sphere.

We still have the transcripts of those lectures, although Galileo later tried to conceal their existence. There is no hint in them that he must

* You can determine which level of Hell you personally belong to by taking a lighthearted Internet test at: http://www.4degreez.com/misc/dante-inferno-test.mv. The best company from my admittedly heretical point of view is probably to be found among the heretics at level 6, including Galileo himself, who shared my taste for scurrilous and sometimes scatological poetry that was far removed from that of Dante.

have had his tongue firmly in his cheek when he described his calculations, which began with working out the size of Lucifer (Satan).

Dante placed Lucifer at the deepest level of Hell, entombed in ice from his navel (which marked the exact center of the earth) up to his breastbone. Just where Lucifer got his navel from must remain a matter for conjecture, and the idea that there is ice at the center of the earth also hardly stands up to modern scientific scrutiny. The method that Galileo used to work out the size of Lucifer involved scaling up the proportions described by Dante from a succession of smaller objects. This was a perfectly reasonable thing to do (mathematically speaking, at any rate), and only later did Galileo realize that he had made a (literally) gigantic mistake in applying the same scaling approach to calculate the dimensions of Hell's roof.

Galileo started by calculating the height of Nimrod, one of the giants who ring the ninth and deepest level of Hell. Here's Dante's description of Nimrod:

To me his face appeared as long and full
As the bronze pinecone of St. Peter's at Rome
With all his other bones proportional.

The bronze pinecone that Dante was talking about now resides in a courtyard inside the Vatican museum and is some 11 feet tall. Based on the fact that most of us are about eight times taller than the length of our heads, Galileo could use simple proportion to work out that Nimrod must have been $8 \times 11 = 88$ feet tall.

That was nothing compared to Lucifer, though. According to Dante, Lucifer was so large that one of his arms exceeded the giant's in size by more (in proportion) than a full giant exceeded Dante himself. Dante was 6 feet tall, so one of Lucifer's arms must have been more than $(88/6) \times 88 = 1{,}290$ feet long. The length of our arms is generally about one-third of our height, so using the same scaling principle, Galileo could work out that Lucifer was at least 1,290 × 3 = 3,870 feet high, or three-quarters of a mile—some 50 percent higher than the world's currently tallest building.

So far, Galileo was on relatively firm ground (figuratively speaking). His use of simple proportion to calculate the size of Hell's deepest denizens only worked, however, for a reason that he could not have been aware of—the force of gravity on a body at the earth's center is zero, so Nimrod and Lucifer would have been effectively weightless.

Nimrod and Lucifer could never have escaped to the earth's surface, though, because (as Galileo later realized) bones have to become *dis*proportionately thicker as they get longer in order to support the weight of their owner. This sets a limit to the size to which living organisms can grow in a gravitational field. Nimrod would have had to be virtually all bone to support his own weight at the earth's surface, and a three-quarters-of-a-mile-high walking, talking Lucifer would be frankly impossible.

Unaware at the time of the disproportionate scaling required for objects to be able to support their own weight, Galileo blithely continued to apply the principle of simple proportion to calculate the thickness of the roof. Here he decided to scale up the dimensions of Brunelleschi's famous self-supporting dome in Florence Cathedral, which is 147 feet wide at the base, yet only 13 feet thick on average. A "simple proportion" scaling-up from a width of 147 feet to 3,245 miles gives a proportional thickness increase from 13 feet to 287 miles, so Galileo estimated that a thickness of 400 miles should provide a sufficient safety factor to guard against collapse. He was shortly to realize that his calculation was literally miles out.

His audience was convinced, however, and he was soon offered a lectureship in mathematics at the University of Pisa. Only after he had taken up the position did he realize that he had made a huge error: The roof of Hell would have to be very much thicker in proportion to support its own weight. The question was, how much thicker?

Once Galileo realized his mistake, he did what a lot of us might have been tempted to do under similar circumstances—he shut up about it and worked on the problem in private to rectify the error before someone else picked it up. Things moved slowly in those days, and fifty years passed before he published the first step toward the correct scaling method in a book called *Dialogues Concerning Two New Sciences*, written while he was under house arrest following his conviction for heresy. Fortunately for posterity, his friends were able to smuggle the book out of Italy, and it was eventually published in Holland by Elsevier Press, which is still active as a scientific publisher.

Dialogues is a sensational book. It laid the groundwork for much of modern physics and engineering, describing the motion of projectiles, a method for measuring the speed of light, and even some scientific party tricks. Above all, it gave the first rule that lets us work out how we can scale up a structure without it suddenly collapsing. The value of that rule may be gleaned from the fact that today's engineers and architects still use it as a rule of thumb for scaling up beamlike structures.

That rule is Galileo's famous "square-cube law." He notes that if you scale up a structure like a beam, a bone, or a body by the same factor in all directions (as he did for the denizens of Hell), then the weight increases in proportion to the *cube* of the scaling factor, but the strength only increases with the area of the cross-section, which is proportional to the *square* of the scaling factor. This means that the cross-section must increase disproportionately for the structure to maintain its strength (see Figure 3.1).

Galileo's diagram of a bone "whose natural length has been increased three times and whose thickness has been multiplied until, for a correspondingly large animal, it would perform the same function which the small bone performs for its small animal." Galileo says that

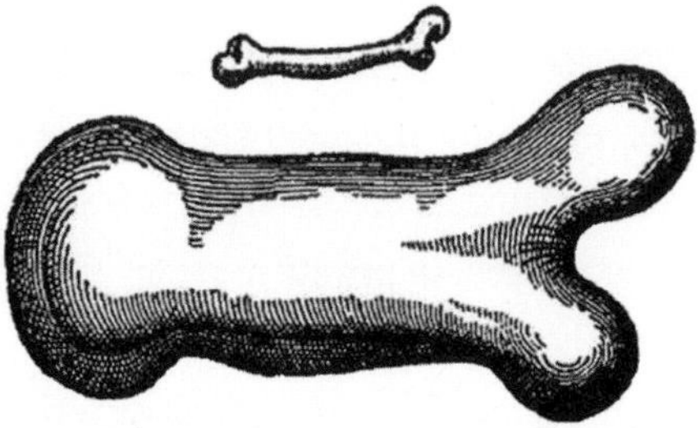

Figure 3.1. The square-cube law: three times the length means five times the diameter. Source: http://galileo.phys.virginia.edu/classes/109N/tns_draft/images/tnsFig027_300.jpg

"from the figures here shown you can see how out of proportion the enlarged bone appears. Clearly then if one wishes to maintain in a great giant the same proportion of limb as that found in an ordinary man he must either find a harder and stronger material for making the bone or . . . admit a diminution in strength."

Galileo realized that the thickness of an unsupported dome would also need to be scaled up disproportionately, but the mathematics of the time was not sufficiently advanced to let him do the calculation. In fact, it wasn't until the 1890s that the correct scaling law emerged—the thickness has to increase with the *square* of the span to maintain the same strength.

The actual thickness depends on the material. For reinforced concrete (today's material of choice), a dome spanning 147 feet need only be 8 inches (two-thirds of a foot) thick. So one spanning 3,245 miles needs to be $(((3{,}245 \times 5{,}280)/147)^2 \times (2/3)) / 5{,}280 = 1{,}668{,}811$ miles thick. Oops!

Early medieval cathedral builders did not even have Galileo's square-cube law to go by. They blithely scaled up columns, buttresses, and arches in the sort of simple proportion that Galileo originally suggested for the roof of Hell—except that this was a race to get to Heaven, with the builder of the tallest steeple winning. Or not winning in many cases, because some 17 percent of medieval cathedrals collapsed soon after they were built.

One of the most spectacular collapses was that of the choir vault in Beauvais Cathedral, just twelve years after the building had been completed in 1272. It was the tallest Gothic cathedral ever built, and in their anxiety to beat the competition, its French builders had made the buttresses *thinner* as well as taller! Various reasons have been suggested for the cause of the ultimate collapse, but the best guess seems to be that the buttresses were so thin that the resulting structure was simply able to flex and either broke directly or vibrated itself to pieces in a high wind. (This was not the only disaster to befall Beauvais Cathedral. A 291-foot crossing tower that was added in 1569, complete with wooden steeple, also collapsed just four years later.)

To make further progress—and develop new scaling laws that would help to prevent such calamities—scientists and engineers needed to understand *why* scaling laws work. Galileo came perilously close to the answer in *Dialogues Concerning Two New Sciences*, but he never quite got there. Amazingly (especially for people who see scientists as omniscient geniuses who could never miss the blindingly obvious), scientists missed the blindingly obvious answer for the next two hundred years.

The answer eventually came in the early nineteenth century with the concept of *stress*, and the recognition that catastrophic structural breakdowns are initiated at points where the stresses have become too high for the material to cope with. We now know that Galileo's scaling law works because it keeps the stresses in a structure at an approximately constant level when the structure is scaled up. But there is a lot more to it than that, and much water would pass under the bridge (or over it if the bridge had collapsed) before scientists truly understood what stress is all about, and how concentrations of stress can lead to sudden and disastrous collapse.

PART 2
HOW DISASTERS HAPPEN

4

The Stress of It All

Beautiful Railway Bridge of the Silv'ry Tay!
Alas! I am very sorry to say
That ninety lives have been taken away
On the last Sabbath day of 1879
Which will be remember'd for a very long time.

—William Topaz McGonagall

William McGonagall was a Scottish street poet who could reduce any disaster to doggerel. Hailed by *Punch* magazine as "the greatest Bad Verse Writer of his age," he demonstrated his talent to the full in "The Tay Bridge Disaster," an epic poem that describes the collapse of the recently constructed Tay Railway Bridge. The poem continues:

Twas about seven o'clock at night,
And the wind it blew with all its might,
And the rain came pouring down,
And the dark clouds seem'd to frown,
And the Demon of the air seem'd to say—
"I'll blow down the Bridge of Tay."

Which the wind did, carrying a train containing seventy-five passengers (not ninety, which we must count as poetic license) into the freezing waters of the Tay estuary during a force-10 gale.

The problem, revealed by the subsequent court of inquiry, was "the insufficiency of the cross bracing and its fastenings to sustain the force of the gale." McGonagall had the answer:

I must now conclude my lay
By telling the world fearlessly without the least dismay,
That your central girders would not have given way,
At least many sensible men do say,
Had they been supported on each side with buttresses,
At least many sensible men confesses,
For the stronger we our houses do build,
The less chance we have of being killed.

Buttresses were a medieval answer to the problem of stresses caused by high winds, poor foundations, and superstructural overloading on the walls of cathedrals and other tall buildings. They prevented the walls from falling over by providing support from the side, as in the beautiful flying buttresses of Notre Dame Cathedral (Figure 4.1, left) and the rather less beautiful buttress that supports my sagging three-hundred-year-old garden wall (Figure 4.1, right).

Buttresses provide one answer, but they are heavy, massive, and often expensive solutions to a problem that modern engineers usually tackle in a very different way—by designing lighter structures that can flex slightly rather than break, and by minimizing the forces generated by the wind through a more open construction. Gustav Eiffel pioneered this approach with the Eiffel Tower, completed in 1889. The structure flexes by two to three inches at the top in normal winds.

Disapproval of Eiffel's design, mostly on aesthetic grounds, was widespread. The novelist Guy de Maupassant was one such critic, but he still reputedly had lunch every day in the tower's restaurant. When asked why, he replied that it was the only place in Paris from which he could not see the Eiffel Tower.

Figure 4.1. Elegant "flying" buttresses supporting a wall of Notre Dame Cathedral in Paris. Photograph copyright © Jean Lemoine (left). Garden wall buttress. Photograph by Len Fisher (right).

McGonagall would also have disapproved, because he saw stern solidity as the answer to the problem of wind-induced stresses. When the Tay Bridge was replaced, he expressed his views in a fresh poem, with just one word added to the first line of the earlier one:

Beautiful new railway bridge of the Silvery Tay,
With thy beautiful side-screens along your railway,
Which will be a great protection on a windy day,
So as the railway carriages won't be blown away. . . .

Thy structure to my eye seems strong and grand,
And the workmanship most skilfully planned;
And I hope the designers, Messrs Barlow and Arrol,
will prosper for many a day
For erecting thee across the beautiful Tay.

Figure 4.2. Forth Rail Bridge. Photographed by an anonymous photographer shortly after its completion in 1890.

The stern solidity of Messrs. Barlow and Arrol, though, was about to be replaced by an almost frivolous lightness that "The Great McGonagall," as he became known, would never have approved of.

One of its earliest manifestations was a sort of horizontal Eiffel Tower—the Scottish Forth Rail Bridge (Figure 4.2), built by the British engineer Benjamin Baker at around the same time that Eiffel was designing and constructing his famous tower.

Like Eiffel's tower, Baker's design was based on a lacelike steel framework in which forces of tension and compression were carefully balanced. But Baker nearly didn't get to design the bridge. The original contractor was Thomas Bouch, designer and builder of the ill-fated Tay Bridge. When that bridge collapsed, his contract was rapidly terminated, and Baker produced his new and revolutionary design.

ALL YOU EVER WANTED TO KNOW ABOUT STRESS

Reality is the leading cause of stress!

Lily Tomlin (as "Trudy the Bag Lady")

Fittingly, Gustav Eiffel was in the crowd at the opening of the new bridge. Baker and Eiffel both recognized the importance of wind-generated stresses and took great pains to estimate them. With no computers to help them in the very complicated calculations (many of which are quite simply impossible without the aid of computers), they had just two aids: their engineering intuition and measurement in the form of stress tests. Baker spent virtually every day on the Forth Bridge while it was being built, measuring the actual wind stresses along its beams and girders. Eiffel went even further: He built a wind tunnel at the base of his tower to carry out the tests.

Even with today's sophisticated computers and mathematical stress models, engineers still use practical stress tests. One of the most spectacular examples is a loaded airplane suspended by its wingtips on huge "jacks" designed for the purpose. Every time you fly, the wings of your airplane are undergoing the same sort of test. Just have a look out the window and see how far the tips bend up. Or, if you are a nervous flyer, concentrate on the airline magazine instead.

Engineers are not the only people who use stress tests. Cardiologists use treadmill stress tests to monitor our susceptibility to heart disease. Psychologists use a psycho-social stress test (called the Trier test) in which moderate to intense emotional stresses are induced under controlled laboratory conditions to test the efficacy of psychotherapeutic interventions. Even the major banks of the world are now undergoing stress tests, designed to measure their long-term stability when placed under financial pressure. Let's hope that these tests perform their intended function of helping the world avoid a repeat of the 2009 credit crunch.

All of these tests purport to measure *stress*. Unfortunately, the word does not necessarily mean the same thing in these different contexts. Many of us think that we understand intuitively what stress means—

and many of us would be wrong, because its meaning in everyday speech can be quite different from what it means for scientists and engineers. The difference can sometimes cause considerable confusion.

The everyday meaning comes from biology and psychology, where stress is often seen as a *response* to a situation. “Reality is the leading cause of stress,” declares Trudy the Bag Lady. “All my stress comes from people not playing the game of life by my rules!” screams the subject in a cartoon that adorns my study door.

Engineers and physical scientists, however, view stress as being *applied* to something. The precise physical definition, originated by the French mathematician Augustin-Louis Cauchy in 1822, is that stress is the *force* (or *load*) applied to something, divided by the *area* over which it is applied. It is tempting to think that Cauchy was inspired to produce this definition following his father’s narrow escape from the guillotine during the French Revolution: The sharp blade of a guillotine is specifically designed to apply as high a stress as possible by concentrating the weight of the blade on a very small area of the neck.

There are plenty of less grisly illustrations. When a woman wearing stiletto heels accidentally steps on my foot, for example, the stress being applied depends on her weight, and inversely on the area of the heel of the shoe. When the wind whistled through the girders of the Tay Bridge, the force it exerted on those girders was transmitted to the small areas of contact between the girders, thus magnifying the stress to the point where the joints gave way.

When real materials are subjected to stress, their response is to deform—eventually up to a breaking point if the stress becomes high enough. The fractional deformation—such as the increase in length of a stretched spring divided by its original length; the fractional compression of a bone in my foot; or the fractional bending of a bridge girder in a high wind—is called *strain.* A lot of confusion could be avoided if we used the term "strain" when we talk about our response to a stressful situation.

GALILEO'S STRESS

Galileo produced one of the first descriptions of how physical stresses can do damage in *Dialogues Concerning Two New Sciences*, where he takes the part of the wise Salviati addressing an open-mouthed and admiring Simplicio:

> SALVIATI: A large marble column was laid out so that its two ends rested each upon a piece of beam; a little later it occurred to a mechanic that, in order to be doubly sure of its not breaking in the middle by its own weight, it would be wise to lay a third support midway; this seemed to all an excellent idea; but the sequel showed that it was quite the opposite, for not many months passed before the column was found cracked and broken exactly above the new middle support.
>
> SIMPLICIO: A very remarkable and thoroughly unexpected accident, especially if caused by placing that new support in the middle.
>
> SALVIATI: Surely this is the explanation, and the moment the cause is known our surprise vanishes; for when the two pieces of the column were placed on level ground it was observed that one of the end beams had, after a long while, become decayed and sunken, but that the middle one remained hard and strong, thus causing one half of the column to project in the air without any support. Under these circumstances the body therefore behaved differently from what it would have done if supported only upon the first beams; because no matter how much they might have sunken the column would have gone with them.

In other words, there was a very high stress at the center of the beam once the support at the far end had been lost. Galileo did not put it in these words, because the concept of stress had not yet been discovered. The key insight, due to Cauchy, was that stress can vary continuously throughout a structure, with some points being under much higher stress than others. Galileo may have missed the point, but his imaginative example carries a clear implication for our modern-day minds: To find out where catastrophic failures are most likely to originate, look for the points in a structure where the stress is highest—that is, where the forces are being concentrated over a small area.

ON BEING UNDER STRESS

Cauchy's equations for stress can get very complicated, but it doesn't need any mathematics at all to visualize the major stresses in simple structures, such as the seesaw illustrated in Figure 4.3.

As the diagram shows, the fibers at the top are stretched the most, but the fibers at the bottom are actually being *compressed.* If you find this difficult to believe, think about what happens when you try to bend an object like a pipe or a straw. There is usually a "crumple zone" on the inside of the bend (Figure 4.4), caused by the compressive forces on that side.

Galileo missed all this because he focused on the forces that the beam was experiencing from the *outside* and assumed that the beam itself was rigid. In fact, marble beams also bend (although much less than wooden ones). The lines of tension and compression in a marble beam are essentially the same as those in a wooden beam—in fact, a beam made of *any* solid material will experience a similar distribution of stresses when put under the same load. Leonardo da Vinci recognized this fact a hundred years before Galileo even started to think about the problem. Unfortunately, the notebook where Leonardo recorded his ideas was not found until 1967, buried in the collections of the National Library of Spain at Madrid.

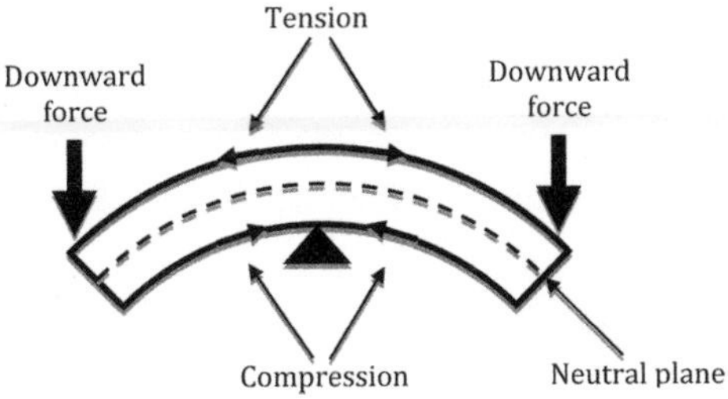

Figure 4.3. Stresses in a seesaw.

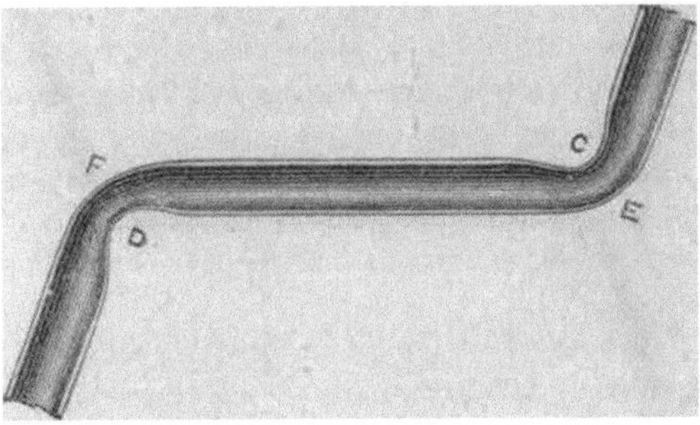

Figure 4.4. Bent pipe showing "crumple zones" (C, D) and zones that have been stretched under tension (E, F). From S. Stephens Hellyer, *Principles and Practice of Plumbing* (1891).

The materials that we think of as rigid are simply those that deform very little under a load. If they can recover their shape after the load is removed, scientists call them "elastic"—a term that applies to glass and steel as much as it applies to rubber bands. My father was fond of telling me when I was a child that glass is almost perfectly elastic, in the sense that it recovers its original shape almost perfectly after it has been bent and the bending force has then been removed. When I later became a scientist, I designed a piece of apparatus that took advantage of this property by using fine glass springs to measure the tiny forces between adhering biological cells. I like to think that he would have approved.

You can't bend thicker glass very far, though, before it breaks. This is because it has reached its *elastic limit*, which is the fractional deformation that it can undergo before it no longer withstands the stress.

The person who discovered the relationship between deformation and force was one of my all-time scientific heroes—the English scientist Robert Hooke, who worked it out in 1676. We know from contemporary descriptions that he had an irritable disposition and that his hair hung in disheveled locks over his haggard countenance. We don't really know what he looked like, though, because he was reputedly so ugly that he refused to have any portraits done.

As well as having these disadvantages, Hooke was born poor and was forced to pay his way through Oxford University by acting as a servant to a richer student. He rapidly achieved a reputation as a brilliant and innovative experimentalist, to the extent that he was offered a position as "curator of experiments" at the newly formed Royal Society of London.

It was a position that few people would seriously consider these days. For a start, it was unpaid. The society also required that Hooke make "three or four considerable experiments"—that is, new discoveries—each *week*! It is no wonder that he had a haggard countenance, especially since he was also required to demonstrate his discoveries to the Fellows of the Society every Friday evening. Amazingly, Hooke managed it, making a huge number of new discoveries in the course of his life—perhaps more than anyone else in history.

One of those discoveries was his law of elasticity, which he published as the anagram *ceiiinosssttuu* because he wanted to claim priority for it without actually telling anyone what it was. When he revealed its meaning two years later, it turned out to be *Ut tensio, sic uis*—"As the extension, so the force."

In other words, when a force is applied to an elastic solid, the solid deforms in proportion to the applied force, no matter whether the solid is being bent, twisted, stretched, or compressed. When you sit on a chair, for example, the frame bends slightly. If someone who is 20 percent heavier than you sits on the same chair, the frame bends 20 percent more. That's Hooke's law in action.

It would be another 140 years before the brilliant Thomas Young realized that Hooke's law could be improved on by considering the stresses and strains experienced by the material rather than the loads and deflections in the structure of which the material was made. In his *Course of Lectures on Natural Philosophy and the Mechanical Arts*, delivered at London's Royal Institution in 1807, Young put Hooke's law in its modern-day form:

Stress/strain = constant

The name for that "constant" is now *Young's modulus*. It is a measure of the flexibility of the material, expressed in units of pressure because *stress* (that is, force per unit area) is expressed as a pressure, and *strain* is just a ration of two lengths and is therefore unitless.

The higher the Young's modulus, the stiffer the material. For rubber, it is about 1,000 Pascals (Pa). For glass and aluminum, it is 10 million Pa, while steel is 30 million Pa and diamond is 170 million Pa.

A FRAGILE WORLD

You don't need to be a mathematician to see from Young's equation (and also from common sense) that the strain, or deformation, is going to be greatest where the stress is greatest. In the case of a wooden seesaw, for example, the point of maximum tensile stress is in the center of the upper surface. This is where the wood fibers are being stretched the most, and this is where a break will occur if the tension becomes too great, as can happen when two overweight parents decide to try the seesaw their children have been playing on (Figure 4.5, bottom).

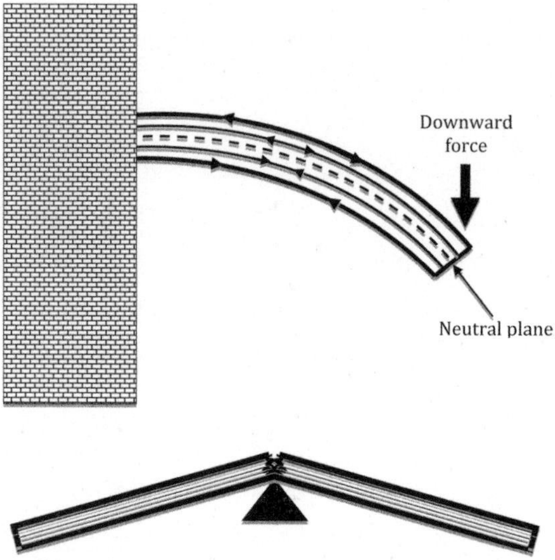

Figure 4.5. Stresses in cantilever beam protruding from a wall (top). Note that the stress varies along the length as well as from top to bottom. So while the entire upper surface is under stress, the stress along that surface is highest in the middle and lowest at the ends. The bottom image shows the place where a seesaw breaks when it is subjected to too much stress.

Why is the stress greatest in the middle? The answer is that the stress along the beam depends on what engineers call the *bending moment*, which is a measure of how strongly the force on the beam is trying to bend it at that particular point. Galileo's calculations showed that the bending moment is highest at the supported end of a cantilever beam protruding from a wall (Figure 4.5, top), and a cantilever beam is just half a seesaw from the point of view of an engineer.

The value of using the concept of stress becomes apparent when we realize that stresses can vary from point to point throughout a structure and that they have a *direction*—that is, they tend to stretch the material of the structure at that point (*tensile stresses*), compress it (*compressive stresses*), or twist it (*torsional stresses*). These can have

Figure 4.6. The Möhne dam in Germany's Ruhr Valley the day after the "Dambuster" raids. Photograph taken by Flying Officer Jerry Fray of No. 542 Squadron from his Spitfire PR-IX on May 17, 1943.

very different effects. When a friend of mine took her first parachute jump and landed with one foot in a rabbit hole, the bones of her leg were able to support the compressive stresses, but the twisting stresses, generated when her body rotated but her foot couldn't, produced a "corkscrew" fracture so severe that it took her years to recover.

The two most common measures of the strength of a hard material like bone or wood are the *tensile strength*—the stress required to pull it apart—and the *compressive strength*—the stress required to crush it. These values can be very different for the same material. The success of the "Dambuster" raids of World War II, for example, relied on the fact that concrete is very strong in compression but comparatively weak in tension. Because concrete is strong in compression, it can support its own weight in a huge structure like a dam, but because of its low tensile strength, the shock wave from an exploding bomb generated tensile forces within the concrete that were strong enough to tear the dam apart (Figure 4.6).

The dam builders could hardly have expected their structure to be subjected to the stresses generated by a bouncing bomb. If they had, they might have considered building reinforcing steel rods into the structure. Steel is very strong in tension—it will stretch, but not break—and it might have saved the dams.

Reinforced concrete is one of the many materials that engineers now use to bear the expected stresses in structures like bridges and buildings. Because those stresses can vary from point to point, it is important to work out where the points of maximum stress might be, since these are the prime locations for incipient disasters.

Computer modeling is now the method of choice. Its success can be gauged by the fact that modern buildings seldom fall down, even in earthquakes. (It is older buildings and poorly designed buildings that most frequently collapse.)

In the days before computers, engineers had to use other means to predict the distribution of stresses. One of the most unusual—and most effective—ways was to measure the shape of soap bubbles formed across appropriately shaped holes in metal plates. Some of the first estimates of the stresses and torques in a twisted airplane wing spar were made in this way during World War I.

The man who came up with the brilliantly simple soap bubble idea was the British aircraft engineer Alan Griffith, who threw a scare into the world of engineers when he showed that finding the points of maximum stress in a structure was only a start when it comes to predicting disastrous failure.

Griffith showed that stress becomes *concentrated* at the tips of cracks, which can be triggered by a defect such as a scratch. He worked out that:

$$\text{Stress concentration factor} \approx \sqrt{([\text{crack length}]/[\text{tip radius}])}$$

Figure 4.7. The USS *Schenectady* after it split in half. From U.S. Government Printing Office, *Board of Investigation to Inquire into the Design and Methods of Construction of Welded Steel Merchant Vessels* (1947).

It doesn't look like a terrifying equation, but it certainly worried engineers, because it shows that the sharper the crack tip, the higher the stress concentration. Worse still, the *longer* the crack, the higher the stress concentration. This means that, once a crack reaches a critical length, it will grow spontaneously and tear a structure apart, as happened to the Liberty ship *Schenectady* in 1943. Moored in calm water at the time, the ship suddenly split in half (Figure 4.7) with a report that could be heard a mile away.

The world is full of cracks and scratches. So why hasn't the world torn itself apart like the USS *Schenectady?*

One part of the answer is that it has. Underground caverns collapse. Large chunks of cliff suddenly crack off and fall into the sea when cracks in the cliffs have grown to a critical size. Something similar happens to glaciers when they "calve." Such processes are happening all the time when the brittle materials of our world are put under tension. Luckily, the vast bulk of the earth's solid material is under compression, and cracks happen only in materials under tension.

Primitive man took advantage of this property of brittle materials by turning them into sharp tools. Flint was a favorite material: One tap could start a crack that would grow until a shard flaked off, leaving a sharp-edged piece that could be used as an ax or a knife.

When metals were discovered, it proved much more difficult to give them sharp edges because metals such as copper and tin (the earliest to be discovered) do not simply crack but also *flow* when they are put under stress. In modern terminology, they are *ductile* and thus less susceptible to catastrophic failure. These are the sorts of material that modern engineers prefer to work with!

Much of the earth's solid matter is a composite of many different materials. Some of these are brittle, but others are softer and more malleable. These composites act as *crack-stoppers*—soft components that yield and absorb energy when a crack reaches them so that a crack is stopped in its tracks.

Wood, for example, contains an ample supply of such materials in the form of gums, resins, and other natural polymers, which is why trees generally don't crack and fall over when the wind blows. If the wind gets too strong, however, some trees *will* crack and fall when the stresses at the crack tips get too high for the crack-stoppers to cope with.

Scientists and engineers have learned the lessons of nature, and many modern composite materials are deliberately loaded with crack-stoppers to inhibit the process of "brittle failure" that Griffith's theory warns us about. Smart "self-healing" materials are also being developed, where the occurrence of a crack initiates a chemical reaction that repairs the material. Some plastics, for example, have microcapsules embedded within them that break open when a crack forms in the vicinity. Some of these contain the original "monomer" from which the polymeric plastic was formed, while others contain a "hardener." When these release their contents, they react chemically (much as do the components of a two-part glue like Araldite) to form new polymeric material that fills the crack.

Another approach to preventing brittle failure lies in surface treatment of the material. The toughened glass in your car windshield and in many store windows, for example, avoids the brittle-failure problem by being treated so that the surface material on both sides is actually under compression. Cracks can't start until the glass is bent so far that the compression on one side turns to tension.

Another way to avoid the brittle-failure problem is to choose materials in which the critical crack length is so long that cracks can be spotted well before they reach the critical length. Mild steel is a case in point. For the stresses that it experiences in most applications, the critical crack length is at least three feet! Some 20 percent of ships now at sea have cracks somewhere near this length, and occasionally one does share the fate of the *Schenectady.*

The dangerous cracks that can be really hard to spot are those that are long but very narrow. Their presence can be detected by X-ray or magnetic examination, but detecting them does not always require such complicated and expensive methods. One cheap alternative that can even be used by the home hobbyist is the commercially available *dye penetrant detector test.* After covering the cleaned part in a dye solution, wash it until no dye is visible. Then spray on a fine powder that sucks out the dye from any cracks and reacts with it to produce a clearly visible pink line.

Another way to test for cracks in hard materials such as metal and glass is simply to tap the material and listen! This goes back to the days of bell casting and continues to the present day in such practices as tapping cast-iron train wheels with a small hammer to check for cracks—even a small crack can alter the tone.

Even with the availability of such tests and the knowledge that has accumulated about the dangers of stress concentration, it still remains a major source of catastrophic collapse. As well as tearing ships, bridges, and airplanes apart, stress concentration also plays a role in natural disasters such as earthquakes, tsunamis, and exploding volcanoes. These events are often much more complicated than I can explain here in this brief outline. Some materials, for example, will flow if a stress is applied slowly enough, but snap in a brittle fashion if the stress is applied more quickly. Just think of chocolate or toffee!

Runaway cracks are driven by *positive feedback*: an effect feeding on itself in an ongoing cycle. This is not the only way, however, that a runaway process can arise. Three other processes that can contribute are *acceleration past the point of no return*, *the domino effect*, and *chain reactions* (which are sometimes, though not always, driven by positive feedback). In the following chapter, I examine all four of these processes and show how they are a primary cause of runaway disasters—not only in our physical world but also in our economic, personal, and social worlds.

5

Runaway Disaster

Stop the World—I Want to Get Off!

—Title of 1961 musical

INSANE ACCELERATION

Runaway acceleration can be a terrifying experience. When I was a schoolboy, a friend turned up for class white-faced, angry, and with a shoe that kept falling off. He was being driven to school by his father in an ancient sedan when the return spring on the gas pedal broke, as was not infrequent in those days. His father struggled to get the car out of gear, and failed. In the few seconds before he thought to turn the ignition off, he had unwillingly passed two trucks, a turning tractor, a startled cyclist, and a police car.

I suspect that my friend invented the police car. What he didn't invent was his missing shoelace. His father had made him take it out of the shoe, then tied one end to the gas pedal and insisted that his son get in the backseat, lean over, and keep pulling on the other end of the lace so as to act as a human return spring.

It's not always quite so easy to escape from such a situation, especially in movieland. My favorite example is from the 1980 *Blues Brothers* film—an archetypal example that has been copied with variations many times since.

In the original script, the Blues Brothers are being pursued in their "Bluesmobile" (a souped-up 1974 Dodge Monaco sedan) by the members of a country music band called "The Good Ol' Boys" in a Winnebago. Unfortunately for the pursuers, Elwood Blues has put glue ("strong stuff") on both sides of the Winnebago's gas pedal. With the gas pedal glued to the floor, and the driver's foot glued to the pedal, the ever-accelerating Winnebago passes a posse of police cars and the Blues Brothers' own car and reaches a *point of no return* when it leaves the road, smashes through a Rolls-Royce dealership, demolishes a hamburger stand, and crashes into a river.

It is the dialogue, though, that really brings the problem into focus:

> BOB (passenger, screaming): Slow down! Stop this damned thing!
> TUCKER (driver, resigned): I'd be happy to oblige you if I could, but I can't, so I won't.

The dialogue encapsulates the fact that the runaway motion of the Winnebago was governed by Sir Isaac Newton's famous *three laws of motion*:

1. *The Law of Inertia:* Every object moving at a constant speed will keep on moving in the same direction at that speed unless it is acted on by an external force.

So even if the rather slow-witted Tucker had had the wit simply to turn the engine off, the Winnebago would still have kept moving at the same velocity (speed in a given direction) unless he had somehow been able to get a foot on the brake. He couldn't stop by running the Winnebago into a tree because he and his companions (none of whom were wearing seat belts) would simply have kept moving forward at the same speed—and out through the windshield.

2. *The Law of Acceleration:* If a force is acting on an object, its rate of change of momentum is equal to the net applied force.

Momentum = mass × velocity. Presumably the mass of the Winnebago stayed pretty much the same, which means that its velocity just kept on increasing under the force of the engine. An equivalent way of putting this is force = mass × acceleration, or acceleration = force/mass.

3. *The Law of Reaction:* For every action there is an equal and opposite reaction.

The "Good Ol' Boys" would probably have liked to sink through the floor of their vehicle, but they couldn't, because according to the law of reaction, the floor would have been pushing back up with equal force to support their weight.

Newton's laws govern the movement of every physical object in the universe—except for cartoon characters, who obey a set of laws that are entirely their own. The first of those laws, according to the *Harvard Lampoon* humorist Mark O'Donnell, is that "any body suspended in space will remain suspended in space until made aware of its situation." When Daffy Duck steps off a cliff, O'Donnell points out, "he loiters in midair, soliloquizing flippantly, until he chances to look down [when he then plummets to the ground]."*

The laws of cartoon physics, however, are not a reliable guide to anticipating disasters. The market analyst Shawn Andrew of the Reflexivity Capital Group is just one of many who have pointed out the parallel with the behavior of many real-life participants in bull markets with an

---

* I'm not sure why O'Donnell chose Daffy Duck in preference to Wile E. Coyote, the Newton of cartoon physics.

unanticipated disastrous crash just around the corner. The assumption of ever-continuing growth is central to the behavior of these participants. "As long as belief in the assumptions continues," comments Andrew, "the abyss remains unnoticed."

Newton's laws, on the other hand, are an essential weapon in predicting and dealing with many real-life disasters. To take just one example, what should you do if you are trapped in a bedroom on the first or second floor of a burning house?

We read about people "leaping from windows" under these circumstances, as though leaping outward will somehow help to break their downward fall. One of the important points in using Newton's laws is to recognize that movements in mutually perpendicular directions (such as vertical and horizontal) are *totally* independent of each other.* If you jump off a diving board, for example, the arc that you follow results from a combination of constant forward speed (by Newton's first law, you keep moving forward with whatever speed you left the board) and accelerating downward speed (by Newton's second law, the force of gravity just keeps on speeding you up as you plummet).

When you hit the water, the force of the impact is still given by Newton's second law (force = mass × acceleration), only in this case it is *deceleration* that matters. Whether we are talking about acceleration or deceleration, we are still talking about (*change in velocity*) divided by (*time over which the change takes place*). So the force with which we hit is, by Newton's second law:

$$\text{Force of impact} = \frac{\text{mass} \times (\text{change in velocity})}{(\text{time over which change takes place})}$$

* This concept may seem trivial to many readers, but when I was teaching the subject I found that it was one that many beginning students had not cottoned on to, and without which they were having great difficulty in understanding how Newton's laws work in practice.

Hitting the water hurts less than hitting the ground because the time to reduce our downward velocity to zero is very short when we hit the hard ground, but much longer when we slide into the water (so long as we dive rather than belly-flop). As this equation shows, the force of impact is proportionally less.

How can we translate this to the problem of escaping from the bedroom of a burning house? We don't have time to go on a diet so as to reduce our mass, but there are a couple of other things that we can do if we don't have time to make a rope of sheets knotted together to climb down. One is to gather up all of the pillows, blankets, cushions, and other soft objects that we can find and *drop* them from a window so that they form a compact, soft landing spot that will spread out the time over which our velocity is reduced once we hit. The second is to *hang* from the window before dropping onto our landing spot, thus reducing the distance through which we have to fall by the equivalent of our height with our arms raised.

1. RUNNING, JUMPING, AND STANDING STILL: ACCELERATION PAST THE POINT OF NO RETURN

When we put Newton's law of universal gravitation and his laws of motion together, a nice little quirk in the algebra ensures that all objects fall to earth with the same acceleration, regardless of their mass. If Galileo had dropped "The Good Ol' Boys" instead of cannonballs from a tower in his famous experiment, he would have found that each of them accelerated downwards at around 32 feet per second^2. After a mere one and a half seconds, they would have been traveling faster than Usain Bolt when he set the Olympic 100 meter record. If they hit the ground at this stage, it would be like Bolt running flat out into a brick wall as he crossed the finish line.

This simple illustration shows just how quickly the acceleration due to gravity can bring us to speeds where impact with a hard surface can cause serious injury or death. Even a short fall can do it, as witnessed by the fact that hundreds of people die in the United States each year from falling out of bed. If you think that sounds ridiculous, imagine running into a brick wall at just one-third of Bolt's speed, which is still a fairly fast jogging pace. The impact will be the same as if you had fallen out of a three-foot-high bed onto a hard floor.

Gravity can also do damage to runners even in the absence of a brick wall, as shown by the case of the fastest cheese in the West—the West of England, that is.

The cheese in question is a circular, seven-pound, orange-yellow "Double Gloucester" produced each year by Mrs. Diana Smart on her family farm in the village of Churcham. It is the one that features in the famous annual cheese-rolling contest down Cooper's Hill in Gloucestershire, not far from where I live.

Up to fifteen thousand spectators watch as hundreds of foolhardy souls chase the rolling cheese down the hill, which starts off with a gradient of one-in-two and steepens to one-in-one lower down. It's a typical example of English eccentricity, and it seems appropriate that it was a rather eccentric Englishman who proved several hundred years ago that the runners will never be able to catch the cheese and that they are likely to face disaster if they try too hard to do so.

The Englishman was Sir Isaac Newton, who spent a small part of his life working out his laws of motion and of universal gravitation, and a much larger part of it dabbling in alchemy and trying to decipher the prophecies of the Bible, which he believed could only be understood once they came to pass.

His laws of motion, though, provide testable prophecies in advance. One of those prophecies is that, even though the cheese is being rolled

Figure 5.1. Chasing the cheese down Cooper's Hill, May 25, 2009. Photograph © Matt Cardy/Getty Images.

down a hill rather than being dropped over a cliff, it still accelerates so quickly that, after a little more than two seconds, it is already traveling faster than Usain Bolt.

Video records show the prophecy coming true. Once the cheese is released, the competitors don't stand a chance of catching it. But Newton's laws haven't finished with them yet. They too accelerate under the influence of Newton's second law as they run down the hill, unless they can find some way of stopping (Figure 5.1).

Newton's third law can help because, when they put a foot out in front and it hits the ground, the ground pushes back on the foot with equal force. It's just a matter of getting enough of the push parallel to the ground, rather than simply acting upwards to support their weight (Figure 5.2, top).

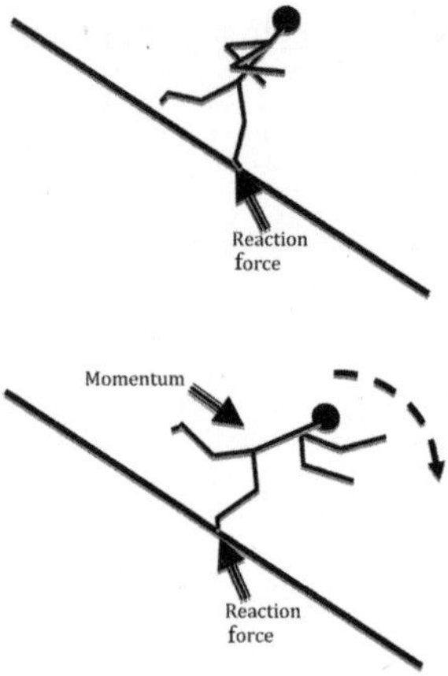

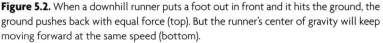

Figure 5.2. When a downhill runner puts a foot out in front and it hits the ground, the ground pushes back with equal force (top). But the runner's center of gravity will keep moving forward at the same speed (bottom).

Unfortunately for most of the competitors, Newton's first law now takes over. The leading foot might hit the ground and stop, but the runner's center of gravity will keep moving forward at the same speed. Unless they do something quickly, the runners will rotate and tumble (Figure 5.2, bottom). In normal running, they can get the other foot out in front in time to stop a fall. Running down a hill, however, as their speed gets higher with each step, the trailing foot needs to come through more and more quickly. The inevitable point of no return comes when it can't be brought through quickly enough and the runner crashes to the ground to become yet another customer for the waiting ambulance men.

2. THE DOMINO EFFECT

The message of the cheese is that Newton's laws are fundamental to that "help, I can't stop!" feeling that comes from unbridled acceleration. Acceleration, however, is not the only way that runaway effects can occur.

Another mechanism is the domino effect,* which is a *knock-on* process: One thing leads to another, which leads to another, and so on, as in the famous Honda TV advertisement that uses car parts instead of dominoes.

Domino toppling is now an art form, with public exhibitions and displays in major galleries. It all started in the 1970s, when Bob Speca, as a college senior, set himself up as a professional domino toppler. He attracted considerable public attention on June 9, 1978, when he tried for a new world record of 100,000 dominoes in a charity promotion for the National Hemophilia Association at New York's Manhattan Center. He had set the dominoes up, and eight-year-old hemophiliac Michael Murphy had provided the initial push. This was when ABC cameraman Manny Alpert dropped his press badge as he leaned over the balcony, starting off a second wave of collapsing dominoes.

Speca prevented a complete catastrophe by quickly removing some dominoes to isolate the damage. Half an hour later, he was the possessor of a new world record—97,500 dominoes. He had also revealed one of the secrets to coping with the domino effect in more serious situations, which is to remove one or more links down the line to stop the effect from propagating.

Believe it or not, the record for domino toppling at the time of writing stands at four *million*, four hundred and ninety-one thousand, eight hundred and sixty-three dominoes. There were, of course, many bifurcations and parallel rows of dominoes in the spectacular arrangement that covered the floor of a large building. As wave after wave of dominoes fell, their toppling initiated other events, including the spinning of records and the unveiling of a picture featuring a naked couple, their backs discreetly facing the photographer.

* Bedouin tribesmen call this effect the *camel's nose*, after the proverb that, once a camel gets its nose inside your tent, the rest will soon follow.

To a scientist's eye, the most fascinating feature of the whole event was the way the wave of toppling proceeded at a more or less constant speed of around two miles per hour along the rows of dominoes. If all the dominoes had been placed in a single line, the event would have taken almost a day to complete, rather than the half-hour or so that it actually took.

When the residents of the German towns of Bremerhaven and Nordenham set up a seventeen-mile-long row of bricks in domino fashion between their two towns to celebrate the opening of a new tunnel, the wave of collapse proceeded at pretty much the same speed as it did along the row of dominoes just described. The bricks were fairly closely spaced, as were the dominoes, which were carefully set with a separation that was only 20 percent of their height.

Theory suggests that the optimum spacing should be more like 40 percent for a maximum rate of toppling. This and other theories also show that the rate of collapse is expected to be constant once the wave gets going. The predictions of these theories agree pretty well with the speeds observed. The theories are only approximate, however, and it is proving extraordinarily difficult to develop a better one. It sometimes seems that we know more about the interior of stars than we do about how to topple a domino!

Why should anyone bother to apply science to such a trivial problem? One answer is that many important advances in science have emerged from studies of simple model problems. In this case, the problem of domino toppling is representative of how cascades of collapse happen in more serious circumstances. Understanding the processes in more detail can help scientists and engineers anticipate such disasters and work out what can be done to prevent them or mitigate the damage.

The domino effect pops up in some surprising places. It has even been implicated in computer password theft. If a hacker gains access to a weakly defended departmental file server and manages to steal some passwords, it often happens that the owners of those passwords are using the same ones in other, more secure systems. A domino effect can result when the hacker uses the stolen passwords to gain access to the more secure systems, potentially revealing additional password files that could lead to the failure of other systems further down the line.

The fairly obvious answer is to use different passwords for different systems, so that the chain is broken—just as Bob Speca did when he removed some key dominoes so as to break a chain of accidental collapse. It's a solution that we can use to help prevent disasters in other areas of life. When we are driving, for example, we can lessen our chances of being involved in a "domino effect" multi-car pileup simply by keeping a sufficient distance from the car in front.

A similar "gap-creating" approach is now being used to anticipate and avoid major industrial accidents, such as the one that occurred in Houston, Texas, in 1989 when a leak in a polyethylene plant caused a vapor cloud explosion. This was bad enough, but the explosion set off others (one in an isobutene storage tank and another in a polyethylene reactor), and these in turn set off a series of seven more explosions in nearby chemical plants.

The problem here was one of having vulnerable plants too close to each other in an industrial complex. If there had been even one gap, the domino effect could have been minimized. New complexes are being designed, and older ones rebuilt, to anticipate the possibility of disasters by incorporating such gaps. The principle is similar to the idea behind the construction of firebreaks in the bushland near my Australian home.

Another way to stop the progressive collapse of a chain of dominoes is to glue one of the intermediary dominoes firmly to the floor. This is the sort of approach that can be of value when it comes to anticipating and preventing the collapse of structures like bridges and buildings. If the engineers had incorporated stronger cross-bracing into the structure of the original Tay Bridge, for example, the failure of one joint might not have precipitated a chain of collapse that sent the whole bridge tumbling into the waters below.

The archetypal example of this sort of collapse is the penultimate scene in the film *Zorba the Greek*. Zorba has built an elaborate structure, consisting of a ramp supported by a series of rickety wooden pylons, to ferry logs down a mountain. The structure wobbles precariously when the first log comes down, but an optimistic Zorba nevertheless calls for a second. When it appears, the first pylon collapses under its weight, so that the second is now bearing all the weight and collapses in its turn, and so on progressively like the collapse of a chain of dominoes.

The Zorba effect in this case leads to nothing more serious than the famous final dance on the beach. In real life, such collapses can have very serious consequences indeed. When the Interstate 35 West bridge across the Mississippi River in Minneapolis suddenly collapsed on August 1, 2007, 13 people died and another 145 were injured. The progression of the collapse was captured on a security camera. The primary cause was later determined to be the fracture of undersized gusset plates that held the girders in place; they had been subjected at the time to an unusually heavy load of construction equipment and material on the bridge above. When the gusset plates failed, the load shifted to nearby points in the structure, which failed in rapid sequence as the bridge spectacularly collapsed and disintegrated.

The progressive collapse of buildings follows a similar pattern. When a gas stove exploded in an eighteenth-floor apartment of the Ronan Point Towers in London in 1968, it blew out a load-bearing wall in the corner of the building, removing the structural support for the four apartments above. They crashed down onto those below, initiating a cascade of collapse down the entire corner of the building. Remarkably, not only did the tenant survive, but when she found another apartment she took her now-repaired gas stove with her!

Domino effects have been implicated in the nationalization of oil industries, the melting of arctic glaciers, the collapse of banks, governments, and economies, and the blowdown of trees in pre-settlement Wisconsin forests. They have also been implicated in the use of hormone replacement therapy for the treatment of ongoing menopausal symptoms, and even in the progression of courtroom legal argument. But domino effects are by no means the worst, or the most damaging, runaway processes that we can encounter.

A wave of collapse along a single line of dominoes happens at a relatively constant speed. If we can get ahead of the wave in time, we have some chance of disrupting it before it can complete its work of disruption. The same general idea seems to work for sequential collapse in real-life situations—so long as we are genuinely concerned with *one* thing following another. If the wave of collapse keeps on dividing, though, we are in a whole different ball game, because we then have a *chain reaction*, and these can develop at frightening speeds.

3. CHAIN REACTION

In a chain reaction, one thing leads to *more* than one other, and each of these leads to more than one other, and so on. Chain reactions are also known as the *snowball effect*, which describes what can happen when you roll a small ball of snow down a snow-covered hillside. As the ball rolls it picks up more snow, gaining more mass and surface area and progressively picking up even more snow and momentum.

The classic example of a chain reaction is a nuclear explosion, where neutrons and other flying fragments from a bursting nucleus break up several nearby nuclei, and the fragments from each of these break several more nuclei, and so on, with energy being released at each step. (To get a feel for the process, you can run your own nuclear chain reaction at http://www.loncapa.org/~mmp/applist/chain/chain.htm.)

Nuclear explosions are not the only chain reactions that have disastrous consequences. The contagious spread of disease, which has caused many more deaths than the atomic bomb, is also a chain reaction—each infected person infects more than one other, and so on.

Nuclear explosions can be harnessed, brought under control, and used to generate power in nuclear reactors when control rods are inserted into the material. The rods absorb a sufficient number of neutrons so that each bursting nucleus affects, on average, *less* than one other nucleus of the fissile material. Energy is still released, but the rate at which it is released is controllable.

Rather similar principles can be used to prevent disease epidemics. Here the "control rods" are people who have been vaccinated and who "absorb" the disease rather than passing it on to others. If a sufficient fraction of people in a population have been vaccinated, the chain reaction of contagion is brought under control, and even simple domino-effect contagion is blocked when the line of infection reaches a vaccinated person. It is even possible to plan the geographical and social distribution of vaccines in an emergency so as to maximize their effectiveness as "control rods."

This picture reveals vaccination to be not only a matter of social choice but also a matter of social responsibility. Getting vaccinated, which protects not only ourselves but others, is an action that we can take as individuals to help prevent the critical transition in the spread of contagious disease from domino effect to chain reaction.

A feature of many chain reactions is the key role played by *proximity*. This is certainly so in the contagious spread of disease, but it can be equally so in the contagious spread of laughter. I have tried this for myself at a crowded party by showing a video of a baby laughing to a person in the corner of the room. When she started laughing, several nearby people also started laughing, and people near them responded by laughing as well. In no time at all, everyone in the room was laughing, but those far from the original source had not the faintest idea of what they were laughing at.

POSITIVE FEEDBACK

The mathematics that describes the progression of a nuclear explosion or the spread of an epidemic through a chain reaction process is similar in principle to that which describes the spread of laughter, the emergence of fads and fashions, and the proliferation of rumors. Such chain reactions underlie many critical transitions, but there is another process that is even more powerful. It is *positive feedback*, a runaway process that is based on the reinforcement of *change*: The reinforcement increases as the change increases, so that an apparently insignificant initial change can grow to become a catastrophically big one. It is nowhere better illustrated than in Douglas Adams's *The Restaurant at the End of the Universe,* a tragicomic science fiction story of the shoe shops on Frogstar World B:

> Many years ago this was a thriving, happy planet—people, cities, shops, a normal world. Except that on the high streets of these cities there were slightly more shoe shops than one might have thought necessary. It's a well-known economic phenomenon but tragic to see it in operation, for the more shoe shops there were, the more shoes they had to make and the worse and more unwearable they became. And the worse they were to wear, the more people had to buy to keep themselves shod, and the more the shops proliferated, until the whole economy of the place passed what I believe is termed the Shoe Event Horizon, and it became no longer economically possible to build anything other than shoe shops. Result—collapse, ruin and famine.

Adams's description shows that positive feedback can have its funny side, but in real life the effects of positive feedback can be far from funny. When I described the process to a chef friend, his first thought was to liken it to the collapse of a soufflé, where a bursting bubble of steam causes nearby bubbles to burst in an ongoing cascade. I pointed out that this was really a chain reaction, but he wasn't listening and his expression had become rather grave. "I've got a better example for you," he said. "It's just like alcoholism." He knew what he was talking about. It turned out that he was a recovered alcoholic, and the meaning of positive feedback to him was that the more you drink, the more you want.

Family arguments provide another example. I well remember one Christmas when I was a child and my parents had produced a game of Monopoly as a family gift. We children quickly lost our money, as did my mother. The game was then between my father and grandfather, who continued to try to bankrupt each other by buying more and more houses and hotels to place on their properties. Each time one of them landed on the other's property, the cries of triumph grew louder, the redness of their faces grew more intense, and the profits were used to buy yet more houses and hotels.

The game went on until three o'clock in the morning and only ended when my mother tipped the board up in exasperation. At least things were quiet from then on—my father and grandfather couldn't bring themselves to talk to each other for the next three days. Positive feedback had generated a critical transition from a state of genial interaction to a state of glowering silence.

Positive feedback can lead to many other critical transitions; in fact, it is the major driving force behind such transitions. Examples include the growth of panic in a crowd; a "run" on a bank; the growth of an avalanche from tiny beginnings; the runaway collapse of a fragile ecosystem; and the evolution of an arms race or an international conflict.

These examples could be multiplied endlessly and are discussed further in the following chapters. But is there anything that we, or nature, can do about them? What we need is a mechanism to keep things in balance—a mechanism that involves *negative feedback*. In the next chapter, I explore how such mechanisms might arise in nature and society. In the subsequent chapters, I reveal how positive and negative feedback processes together act to produce a rich array of possibilities. Apparent stability can often be a prelude to dramatic transitions; we need to learn how to read and heed the warning signs that precede them.

6

The Balance of Nature and the Nature of Balance

Unfortunately the balance of nature decrees that a super-abundance of dreams is paid for by a growing potential for nightmares.

—Peter Ustinov

The man with muttonchop whiskers pulled strongly at the oars of his small wooden boat, thrusting his way through the thick mat of weeds that hindered his progress across the shallow Illinois lake. Even on this cold February day he was perspiring in his black frock coat as he paused occasionally to pull up a handful of plants, observing with interest that they were covered with tiny shells and alive with small crustaceans. Behind the boat trailed a dredging net, designed to catch the more mobile denizens of the lake—the "pumpkinseed," a sunfish whose resplendent colors of orange and yellow were set off with patches of green, scarlet, and purple; a range of other sunfish; two varieties of black bass; the common perch; and a host of others.

The fish were there to feed on the small creatures that lived among the weeds. The man was there because he wanted to understand the relationship between all of the lake's inhabitants. He would later describe it as "an equilibrium of organic life" in which every species played a role. In modern terminology, he viewed the lake as an *ecosystem*. He was one of the first people to think about nature in this way.

The man was Stephen Forbes—a self-taught scientist who had become the official entomologist for the state of Illinois. His pioneering studies of the northern Illinois lakes in the early 1880s would eventually win him fame as a father of modern scientific ecology. He provided one of the first scientific explanations of what has become known as the *balance of nature*—the idea that natural systems have a built-in system of self-regulation whose checks and balances ensure the long-term stability of the system, so long as we don't interfere with it.

According to Forbes, natural systems such as his lakes were stabilized by "a harmonious balance of conflicting interests" where "an order has been evolved which is the best conceivable without a total change in the conditions themselves; an equilibrium has been reached and is steadily maintained that actually accomplishes for all the parties involved the greatest good which the circumstances will at all permit."

Forbes's self-imposed task was to understand how the processes of nature could produce such a balance between different species. How could the community in his little lake maintain its equilibrium, for example, in the face of fierce competition among its members for food, survival, and reproductive success? Why didn't the predators eat up all of their prey and then die out themselves for lack of food?

He saw the answer in the different reproduction rates of the various species. Prey species, he argued, must "produce regularly an excess of individuals for destruction, or else . . . must certainly dwindle and disappear." A predator species, on the other hand, "must not appropriate, on an average, any more than the surplus and excess of individuals upon which it preys, for if it does so, it will regularly diminish its own food supply, and thus indirectly, but surely, exterminate itself."

In other words, predators needed to limit their numbers to suit the available food supply, and the best way for predator and prey to survive simultaneously was for each to adjust its reproduction rate to suit not just its own needs but those of the other.

Forbes was not the first person to think of this argument. That honor goes to the ancient Greek historian Herodotus, who pointed out some two and a half thousand years ago that "timid animals which are a prey to others are all made to produce young abundantly, so that the species may not be entirely eaten up and lost; while savage and noxious creatures are made very unfruitful."

Herodotus believed that the different reproduction rates were the work of a beneficent divine providence that "made" creatures to fit into a predetermined pattern. Forbes was a rationalist and agnostic who did not deny the possibility of a divine providence, but who sought for a more earthly explanation. He found it in the concept of "survival of the fittest" that Charles Darwin had introduced in 1869 in the fifth edition of his book *On the Origin of Species by Means of Natural Selection and the Preservation of Favored Races in the Struggle for Life.*

The phrase "survival of the fittest" was coined by the economist Herbert Spencer after he read the first edition of Darwin's book. Darwin found it to be an appealing metaphor that encapsulated the processes of natural selection, and he took it up with enthusiasm in the later editions of his famous work. He might not have been quite so enthusiastic if he had realized that the term would be open to a gross misinterpretation that persists to this day—the erroneous idea that the strong must inevitably dominate and survive at the expense of the weak.

This misinterpretation of "survival of the fittest" has been used to justify persecution, war, and even genocide. These sickening misuses have nothing to do with what Darwin meant in his own use of the term. If he is watching from the grave, he is probably revolving in it.

Forbes used "survival of the fittest" in Darwin's original sense—as the idea that the individuals most likely to survive and reproduce within a particular environment are those that are best adapted to survive in that environment. Where Forbes showed his originality was in demonstrating that this could account for the "co-adaptation" of different species, and in particular for the observed balance between populations of predator and prey.

Forbes's argument was that the fittest members of a prey species would be those that produced enough young for just a few to survive to reproductive adulthood. Individuals that produced too few would soon be removed from the genetic pool because their young would be eaten before they could grow and reproduce. Individuals that produced too many young would also be at a genetic disadvantage because they would overconsume their food supply and die of starvation before reaching adulthood.

He applied the same arguments in reverse to predators. The fittest individuals would be those that bred enough young to grow and reproduce, but that otherwise kept the number of young to a minimum so that there would be enough food for all.

In terms of modern genetics, the "fittest" individuals, whether predator or prey, would be those that have the best chance of surviving to adulthood, where they could pass their genetic material on to the next generation. Their genetic inheritance would thus eventually come to dominate the species. As Forbes elegantly put it: "The beneficent power of natural selection compels such adjustments of the rates of destruction and of multiplication of the various species as shall best promote [their] common interest."

Forbes presented his arguments in a paper that he read to the Peoria Scientific Association in the evening of Friday, February 25, 1887. Called "The Lake as a Microcosm," the work is now rightly regarded as a classic. To those members of the audience who were still awake at the end of his hour-and-a-half-long talk, Forbes's conclusions must have come as a bombshell.

Many of Forbes's listeners would have had similar Puritan upbringings and would have been taught, like him, to take a holistic view of nature. Now here was a powerful scientific argument in favor of that viewpoint—an argument, moreover, that provided a *mechanism* for the existence of a long-term balance in natural communities. This mechanism was based on the already well-known principle of "negative feedback," although Forbes himself never used the term. At the time of his talk, it was applied almost exclusively to physical systems. It would be another half-century before negative feedback began to assume the central position in biology that it continues to maintain today.

NEGATIVE FEEDBACK: FROM BALL COCKS TO BALANCING TRICKS

Negative feedback is one of the most powerful mechanisms available for maintaining stability. It is a restoring mechanism that works by pushing a system back toward its original state, with a push that gets stronger the further the system has moved away from that state.

Our world is full of practical examples. Driving a car requires the use of negative feedback. When the car starts to drift away from the line that we want it to take, we turn the steering wheel in the opposite direction to bring it back into line. The farther the car deviates from the line, the more we have to turn the wheel to get it back on the line.

One of the few people who never mastered this process was Stephen Forbes himself. According to his son, speaking at Forbes's memorial service, he became locally famous in his hometown of Urbana for a long series of minor mishaps "resulting from the facts that the automobile came late in the period of his physical adaptability, and that he drove to the accompaniment of his intensely concentrated thinking, without knowing that he failed to give the job his whole attention."

Luckily, negative feedback doesn't always need conscious action. Usually it is a built-in property of a system that operates automatically.

One of the earliest examples is the magnetic compass, whose needle swings back after it is disturbed so that it always points to the north. Another example is the centrifugal flyball governor, which was invented in 1788 by James Watt to regulate the speed of his rotary steam engine. This device employed two pivoted rotating flyballs that were flung outward by centrifugal force. As the speed of rotation increased, the flyweights swung farther out and up, operating a steam flow throttling valve that slowed the engine back down.

The thermostat in your home operates on a very simple negative feedback principle. When the temperature drops below the desired value, a sensor in the thermostat sends a message to switch the heating on. When the temperature reaches the desired value, the sensor sends another message to switch the heater off. The same process operates in reverse with the air conditioner cooling cycle.

Thermostats are an obvious example of the negative feedback principle. A rather less obvious example is the ball cock in your flush toilet, which has its origins in a float control for a water clock devised by the Greek inventor Ctesibius around 270 BC. When the cistern is full, the ball-cock arm holds a valve closed to stop water from entering. When you flush the toilet, the water level in the cistern drops and the ball cock floats down with it, allowing water to flow into the cistern by tilting the arm and opening the valve. The more the arm tilts, the wider the valve opens, and the faster the water can flow into the cistern. As the cistern fills, the valve gradually closes as the water level returns to its "equilibrium" position.

Your body also uses negative feedback in a myriad different ways. When you get too hot, you start to perspire more, and the additional perspiration cools your body down as it evaporates. Literally thousands of negative feedback processes are going on inside your body to maintain the stability of everything from your heart rate to the balance of chemicals in your bloodstream.

We learn about negative feedback early in life when we make our first attempts to balance objects on our hands. For instance, if you have ever tried to balance an umbrella, you know that, when it starts to tip, you move your hand sideways until the umbrella starts to tip the other way and comes back to the equilibrium balance position, and eventually you achieve the requisite hand-eye coordination. You underwent a similar learning process at an even earlier stage when you were learning to walk. "Every step," my father later explained to me, "is a fall caught in time."*

Negative feedback need not be such an active process. It is a built-in property of many materials that is described by Hooke's law. Bending, stretching, twisting, or compressing something generates a restoring force that increases with the deformation. So the slight deformation of a chair when we sit on it and the bending of a beam as it supports a bridge or a building are both examples of negative feedback.

STABLE OR UNSTABLE: A TALE OF CHAIRS, TORTOISES, AND THE GÖMBÖC

Negative feedback promotes *stable equilibrium*. One of the messages of this book is that stable equilibrium can be destroyed when runaway processes such as positive feedback begin to take over and sometimes lead to disaster. To anticipate and deal with such disasters, we need to be able to predict the changeover point. Physicists and engineers label that point as *unstable equilibrium*—a point where things could go either way with just a small change of circumstances, as happened when I once leaned too far back on a chair.

---

* When we walk, we are temporarily poised during each step on the end of one leg, tilting to a position of unbalance before we are prevented from falling by the contact of the other foot with the ground—a process of negative feedback that seems to be learned rather than innate.

So long as I didn't lean back too far, I was in a position of stable equilibrium—thanks to negative feedback, which arises from a neat twist to Newton's laws.

My weight was acting downwards through my center of gravity. According to Newton's third law, the floor must have been pushing back up with the same force: The floor would have bent slightly under my weight, and its springiness then generated the required upward force via Hooke's law.

The twist came (literally) because the upward and downward forces were not acting in the same line. They were separated in space to form what scientists call a *couple*—in other words, a pair of parallel forces working in opposite directions that together tend to *rotate* an object. In this case, those forces were acting to return the chair to its equilibrium—an upright state (Figure 6.1a).

The force of rotation is called a *torque*. When my classmates and I learned about torque at school, our physics teacher came up with a memorable mnemonic designed to appeal to our testosterone-driven instincts: "The closer the couple, the less the torque."

As I leaned back farther, the restoring torque became less. Eventually, I reached the position of *unstable equilibrium*, where the upward and downward forces were collinear (Figure 6.1b). Not realizing the danger, I allowed the chair to tilt slightly farther. Disaster! The torque (Figure 6.1c) was now rotating the chair *farther away* from its original equilibrium position. The rate of rotation was also increasing more and more rapidly under the influence of positive feedback—the farther I tilted, the greater the torque became.

All too soon I arrived at a new equilibrium position: flat on my back, waving my arms and legs around like a tortoise that has been turned over (Figure 6.1d). Some tortoises, though, would have had a much better chance than I did of righting themselves with dignity, thanks to

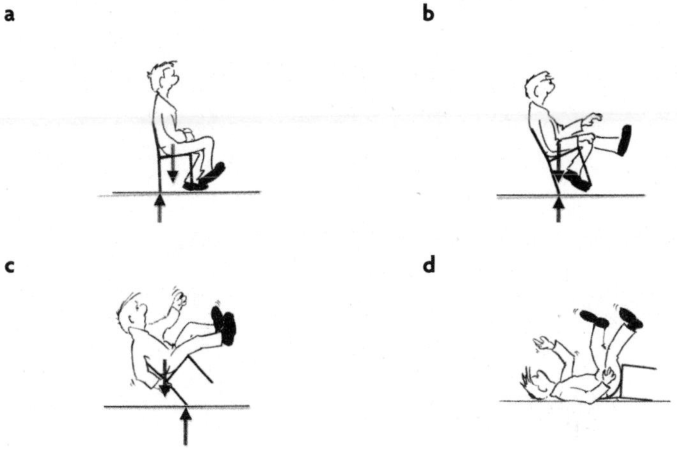

Figure 6.1. Equilibrium (a) stable; (b) unstable; (c) lost; (d) a new equilibrium.

the unusual shape of their shells. One of these is the Indian star tortoise, whose dome-shaped shell is shaped like the recently discovered *gömböc* (pronounced *goem-boets*)—a three-dimensional convex object (Figure 6.2) that can "self-right" from virtually any position to a unique equilibrium position.

If you put a gömböc down on a horizontal surface, it will start wobbling around until it has safely reached that equilibrium position.* The gömböc can self-right because its clever shape ensures that its center of mass is always on the side of the reaction force that produces a couple acting to rotate it toward its equilibrium position. In other words, no matter how you put it down, it will always self-right through a process of negative feedback. The same goes for the tortoise, with its gömböc-shaped shell.

* See the video at http://www.youtube.com/watch?v=pn811yIALPw, which also shows the tortoise self-righting.

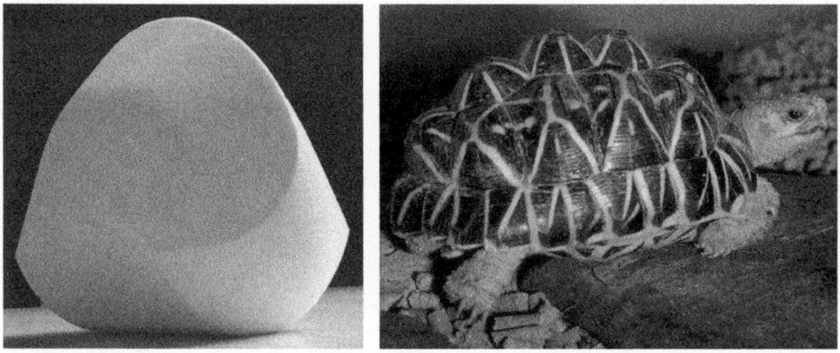

Figure 6.2. The gömböc. Note that the density of the gömböc is uniform throughout; the balance does not depend on having an additional weight inserted at some point (left). Indian star tortoise. Photograph courtesy of James Harding, Department of Zoology, Michigan State University (right).

RIGID BELIEFS

Negative feedback is not the only way to maintain stability. Another approach is to create a rigid structure that simply cannot depart significantly from its normal position, like the strongly buttressed bridge design recommended by "the Great McGonagall." Or like the office chair of one of my scientific colleagues that a group of us glued to the floor while he was on holiday.

Modern engineers create near-rigid structures by using a rod-and-joint framework that maintains a balance between the tensions and compressions in its members. Such a structure, like the Eiffel Tower or Buckminster Fuller's many geodesic domes (Figure 6.3), can fail catastrophically if the stresses become too high, but it is otherwise remarkably stable.

In Chapter 8, I draw some parallels between maintaining the stability of physical structures through rigidity and some societies attempting to maintain their stability by preserving the status quo—an approach that has also been used for economies and ecosystems. One

Figure 6.3. Eiffel Tower, Paris (left). Panoramic view of the geodesic dome structures of the Eden Project, a large-scale environmental complex near St. Austell, Cornwall, United Kingdom. Photograph by Jürgen Matern (right).

of the main pitfalls is that local stresses tend to build up and cracks can appear in unexpected places, as I discovered when I tried to imitate one of the courtroom tricks of the famous defense attorney Clarence Darrow.

Darrow's opponent was addressing a jury in a case that Darrow seemed bound to lose. "During [the] case," said New York State Supreme Justice Louis B. Heller, "the DA, who had an ironclad case against the defendant, began summing up. As the prosecutor began to speak, Darrow lit a cigar into which he had inserted a wire. He took a puff from time to time, but never flicked the ash."

"A gradually lengthening cylinder of ash held the fascinated gaze of the jurors, who were waiting for it to fall off. They concentrated on Darrow's cigar, while the prosecutor talked on."

"Needless to say, the ashes never came off, and the jury's attention was effectively distracted."

When I tried this trick at a party, the ash stayed on the wire, but it broke up into discrete discs as the cylinder got longer, so that the wire was exposed and the trick revealed. I argue later that economies, societies, and ecosystems that use rigid control to maintain stability can suffer a similar sort of breakup. Rigid control is not a foolproof recipe for long-term stability.

UPSIDE-DOWN GRAVITY

One other idea from engineering is that of *driven equilibrium.* Unlike negative feedback, driven equilibrium works even if there is *no* position of stable equilibrium. It is the basis for a modern version of the famous Indian rope trick.

I learned about it when I was presenting a radio series called *How to Find the Sweet Spot.* I had surfed on a tsunami, worked out the best way to hit and kick a ball, and examined building design in earthquake zones. Now I was looking at how flexible objects like chains, ropes, or inverted pendulums could be balanced stably on their tips in apparent defiance of the laws of gravity.

This extraordinary possibility was discovered by a scientist called A. Stephenson in 1908. Little seems to be known about him. He published his analysis in the *Memoirs and Proceedings of the Manchester Literary and Philosophical Society.* The society still exists, and I asked the current secretary if A. Stephenson was related to the famous former member George Stephenson of steam engine fame. Sadly, the answer was no. In fact, the society's records do not list A. Stephenson as having been a member at all.

Stephenson worked out that a chain of linked rods could stay balanced on its end if the support was vibrated up and down very rapidly in a process known as "forced oscillation." The trick is obvious once it

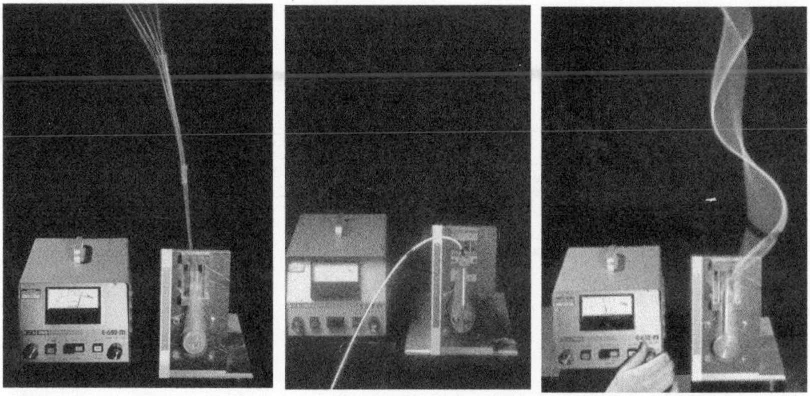

Figure 6.4. Three linked rods maintained in vertical configuration by rapid vertical oscillations at the base. David J. Acheson and Tom Mullin, "Upside-Down Pendulums," *Nature* 366 (1993): 215–216. Wire on stationary base (center). The same wire and base, but with the base oscillating rapidly in the vertical direction to support the wire "balanced" on its end. Tom Mullin et al., "The 'Indian Wire Trick' Via Parametric Excitation: A Comparison Between Theory and Experiment," *Proceedings of the Royal Society of London* A459 (2003): 539–546. All images courtesy of Professor Tom Mullin.

is explained. It has to do with "upside-down gravity"—the vibrating support generates a vertical upward force during that part of the cycle when it is accelerating downwards, and this force more than counteracts the downward force of gravity during that part of the cycle. It was nearly a century before the conditions for driven equilibrium were realized in practice by the Manchester physicist Tom Mullin, whose apparatus for balancing a set of loosely linked rods (Figure 6.4, left) I was "demonstrating" on my radio program.

The genuine Indian rope trick requires that a single flexible object like a rope be "balanced" on its end. Tom later managed this, but with a wire rather than a rope (Figure 6.4, center and right).

"Driven equilibrium" is a clever trick, but fiendishly difficult in practice because of the stringency of the conditions with regard to the amplitude (the distance) of the oscillations, their frequency, and the length and masses of the individual rods. Thus, it finds little practical application in maintaining the stability of large-scale physical structures. I mention it here partly for completeness, partly because it is fun, and partly (as I show in Chapter 8) because I suspect that analogous processes may help us to overcome some of the problems of stability that arise in our complex society.

MAINTAINING THE BALANCE

When it comes to maintaining a balance, negative feedback is usually the best option. But it needs to be quick. When you were learning to balance an umbrella or a broom on your hand, you quickly discovered that it is easier to maintain the balance if you apply small corrections rapidly, rather than applying large corrections more slowly, and usually when it is too late. This is the basis of *control theory*, developed by the Scottish physicist James Clerk Maxwell in 1868 and now ubiquitous in our world through the development of fast electronics and computers to do the controlling.

An everyday example is the cruise control in your car. A sensor tells an inboard computer how fast the car is going. The computer compares this with the speed that you have told it you want to go and uses the difference to calculate how much to vary the fuel flow to the engine. The final step in the loop is the computer's signal to the fuel pump, telling it to speed up or slow down as appropriate. This *feedback control loop* is repeated many times per second to maintain smooth control. Space engineers use similar electronic feedback principles to keep their rockets on course, and manufacturers use it to optimize the performance of their production lines.

Figure 6.5. A chain of Segways. Courtesy of Segway, Inc.

One impressive application of the feedback control loop principle is in the previously "impossible" device known as the Segway—a scooterlike traveling platform that stays upright in apparent defiance of the law of gravity (Figure 6.5).

Scientifically speaking, the Segway is an *inverted pendulum*—that is, a pendulum that swings upside down. It doesn't fall over because gyroscopes sense any departure from the vertical and send a signal to a built-in computer, which completes the negative feedback control loop by instructing motor drives to move the wheels in response.

Segway boss Jim Heselden sadly died when he fell over a cliff in a freak accident while riding the latest model, leading to some truly ridiculous media speculation about the machine running out of control. The plain fact is that it is almost impossible to lose control of a Segway, although George W. Bush managed it. There is a YouTube video of this event, alongside a video in which a chimpanzee was filmed successfully riding one. But it wasn't the relative balancing ability of the two hominids that was at issue. It was the fact that Bush seemed not to have realized that a Segway needs to be switched on if it is to remain upright.

The mathematics and electronics that underlie feedback control loops, such as that utilized by the Segway, are well beyond the scope of this book, but their hidden hand stabilizes many of the processes in our world, from the thermostats that control the temperature in our refrigerators to the amplification of the music that we listen to and the stability of the television signals that we receive. In all cases, the principle is the same—measure a signal, compare it to the one that you want, and use the difference to calculate and apply an appropriate correction.

MICROCOSM

The feedback control loop sounds very similar to the negative feedback principle that Forbes proposed to account for the balance of nature. The big difference is that control theory lets us specify the "equilibrium position" that we want, while the equilibrium position in nature is generated by the feedback process itself.

Forbes saw his lake as a microcosm of human society, which he believed was stabilized by comparable processes of built-in self-regulation. More than a century earlier, the famous Scottish economist Adam Smith had a rather similar idea when he proposed that capitalist economies have their own natural, self-regulating, optimum equilibrium, driven by the "invisible hand" of competition.

Smith even extended his arguments to the concept of democracy and a free society, in which he saw self-interest as the driving force to maintain stability. "Self-seeking on the part of individuals," he said, "will in society so interact with and check itself as to be for the benefit of the social whole . . . [leading to a] delicate balance." The poet Bernard de Mandeville had put the same argument more succinctly nearly two hundred years earlier in a poem called "Fable of the Bees": "Private Vice is Publick Benefit."

These are powerful arguments on the surface. They account for the way complex ecosystems, economies, and societies remain stable for such long periods of time by proposing that negative feedback holds them in check. "Private vice," for example, is no more than control theory applied to life: We act to restore a favorable position whenever we find circumstances taking us away from that position. Adam Smith's hypothetical "invisible hand" also works by negative feedback, with the balance between the lowest prices for the consumer and the highest profits for the producer working like the governor on Watt's steam engine to maintain both at an equilibrium level.

But can these processes act to produce an ideal equilibrium, as Smith and Forbes believed and as some people still seem to believe? Jon Meacham, the editor of *Newsweek*, is one such believer. He put the argument for the "invisible hand" of self-interest acting to produce an ideal equilibrium when he claimed that "'Democracy is a Pesky Thing' . . . [but] history suggests that in the end, after much trial and much error, we usually get it right."

Wrong. History suggests no such thing.

What history *does* suggest is that the evolution of societies and natural communities is a complex process that often proceeds in a series of bursts, with periods of stability interspersed with sudden, dramatic, and usually unforeseen transitions to a different state. The historian Arnold Toynbee compared the evolution of societies to a musical rhythm of three and a half beats to the bar. He noted that, for many civilizations, a period of initial growth was followed by an "event" marking the end of growth, and then a three-and-a-half-beat pattern, lasting some one thousand years, of collapse . . . recovery . . . collapse . . . recovery . . . collapse . . . recovery . . . final collapse (Figure 6.6).

Similar patterns are observed in banking systems, national economies, personal relationships, and world weather patterns, as well

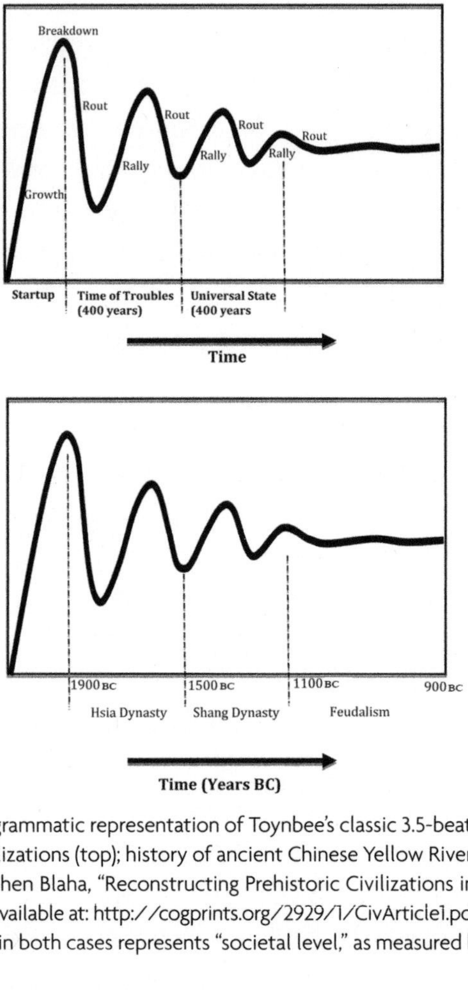

Figure 6.6. Diagrammatic representation of Toynbee's classic 3.5-beat pattern for the evolution of civilizations (top); history of ancient Chinese Yellow River civilization, redrawn from Stephen Blaha, "Reconstructing Prehistoric Civilizations in a New Theory of Civilizations," available at: http://cogprints.org/2929/1/CivArticle1.pdf, p. 19 (bottom). The vertical axis in both cases represents "societal level," as measured by a range of indicators.

as in the evolution and propagation of biological species, not to mention ideas. These are not the sort of patterns that we might expect if negative feedback ruled alone. They are much more the sort of patterns that we might expect if there were a mixture of negative feedback and runaway processes like positive feedback, with negative feedback maintaining an uncertain equilibrium until circumstances reach a critical transition—the point where a runaway process can take over.

A particularly dramatic example of such a critical transition was the explosion of the Icelandic Eyjafjallajökull volcano on April 14, 2010, which brought European air traffic to a standstill. Pressure in the molten underground magma had gradually built up over time, and it had reached a point where only a slight further increase was needed to blow the lid off. This particular lid consisted of a 1,500-foot-thick glacier, whose water combined with the materials of the magma to produce a devastating plume of fine glassy particles that the atmospheric jet stream carried across Europe.

This volcanic explosion was not the only critical transition to affect Iceland and its neighbors in recent times. On October 9, 2008, the value of the Icelandic *króna* collapsed, the national banks of Iceland failed, and Iceland itself became bankrupt. The underlying cause was a gradually increasing public mistrust of the ability of the banks to repay money that had been deposited with them on the promise of high interest rates. The banks had loaned out the money on the basis of optimistic economic forecasts, and without sufficient security. Now they were being called to task, as were banks in other countries during the global economic crisis. The result was an escalation of mistrust and eventually a runaway collapse.

Such critical transitions have strong parallels with those that the ecologist Marten Scheffer and others have found in shallow lakes, where just a small change in circumstances can tip the lake from a clear to a turbid state. The lake cannot immediately be returned to its original clear state simply by reversing the change. There *is* a shift back to the original state, however, if the change is reversed far enough—that is, if a sudden (not gradual) critical transition is initiated in the opposite direction.

A similar sort of cycle happens over and over again in markets of all kinds. Economists call it the "market cycle" and tend to picture it as a relatively smooth process. More often than not it is closer to the *punctuated equilibrium* that the paleontologists Niles Eldredge and Steve Gould have proposed for the process of evolution: long periods of relative stability or slow change interspersed with short periods of sudden and dramatic change.

The seminal 1972 article in which Gould and Eldredge introduced their evocative descriptive term has rightly become famous. Gould was not pleased, however, when I pointed out to him after a talk in London that the novelist G. K. Chesterton beat them to it by nearly one hundred years.

Chesterton used the term in a short Father Brown detective story called "The Strange Crime of John Boulnois," which has striking parallels with Gould's later discovery. The Boulnois of the story is an Oxford philosopher who spotted some "allegedly weak points in Darwinian evolution" and has introduced an alternative theory called "Catastrophism," which has had "some rather faddy fashionableness at Oxford."

In reality, Chesterton knew nothing about science, and he had invented this story just to give his character some verisimilitude. I thought that the parallels with Gould's proper scientific work were rather funny, but Gould did not see the joke. After I had provided him with a copy of Chesterton's story, I heard no more.

Eldredge and Gould's real scientific theory went on to become a major strand of evolutionary thinking. Similar thinking now pervades the social sciences, with "punctuated equilibrium" being invoked to explain phenomena as disparate as organizational transformations, the diffusion of new technologies, and the dynamics of U.S. environmental policy. But how can we explain the frequency of punctuated equilibrium processes in nature and society? Where do the sudden jumps come from?

It doesn't take a genius to tell us that they must arise when runaway processes like positive feedback take over. But how does this happen? How do positive and negative feedback processes interact in complex systems? Most importantly, how can we predict when the runaway processes that so often lead to disasters are about to take over from the stabilizing negative feedback processes that provide a comforting illusion of long-term stability?

The first real steps toward an answer came in the early 1970s, when physicists and mathematicians started to apply their way of thinking to the big problems of biology, such as the processes of balance and collapse in ecosystems. Even in the simplest cases, though, the answer turned out to be not so simple. My attention was drawn to it in a rather odd way when I came across an article, published in the prestigious scientific journal *Nature*, on the ecology of dragons.

PART 3
IMMINENT CATASTROPHES: READING THE SIGNS

7

The Chaotic Ecology of Dragons

Meddle not in the affairs of dragons,
for you are crunchy and taste good with ketchup.

—Anonymous

"The Ecology of Dragons," with its tongue-in-cheek allusions to many contemporary conservation problems, has become a cult classic among ecologists. Its author sought to bring to public attention a remarkable piece of research on how this previously dominant species had disappeared and to add his own theories on how they might have evolved in the first place. When I read it, I was fascinated to find that dragons belonged to the same lineage as centaurs and angels; that they could live for up to ten thousand years; that one was kept captive by Pope St. Sylvester around AD 350 and consumed six thousand people daily; and that the chaos caused by dragons ended in catastrophe (for the dragons) when they became extinct in the eighteenth century.

The author of the article suggested that the dragons might have died out because of commercial overexploitation, possibly for pharmacological purposes. Dragons' heads were particularly in demand because they contained *draconites*—gems that could cure epilepsy, prolong life, convey invincibility, or be used as fireworks. The blood of dragons was also sought after because it could dissolve gold or, in hot climates, be used as a coolant.

Thus, dragons were slaughtered in great numbers. They were a protected species, but only on the island of Rhodes, where the king had enacted legislation in 1345 expressly forbidding any knight to attempt to slay a local dragon. (It has been suggested that this was more for the protection of the knights than the preservation of the dragons.)

An alternative explanation for the dragons' demise was that the supply of medieval knights (the dragons' principal food supply) simply gave out. Without an adequate supply of knights, the dragons starved to death. In other words—and I believe that I am the first to point this out—the dragons died from knight starvation.*

The author of this spoof article turned out to be none other than my former bridge partner, Bob (now Lord) May, who had been a rising star in the physics department of Sydney University when we played bridge together. I was mightily impressed to find that he had now transmogrified himself into a professor of zoology at Princeton University—not on the strength of his knowledge about dragons, but because he was one of the first scientists to show how the mathematical methods used by physicists could fruitfully be applied to the problems of ecology.

His article stimulated me (as it did many others) to find out more. I discovered that he had been attempting to describe mathematically the question that Stephen Forbes had addressed intuitively—what is the long-term outcome of the dynamic interactions between different populations of organisms in an ecosystem? With dragons and their knightly food supply, for example, was it possible to analyze mathe-

---

* There is a widespread popular belief that dragons captured and devoured fair maidens, but this is far from the truth. In reality, as pointed out by Holger Jannasch of the Woods Hole Oceanographic Institute, dragons hardly ever ate the fair maidens they had captured, but used them as bait to attract the knights that were their favorite food.

matically the circumstances in which the knights might have killed off all of the dragons? Or the dragons might have eaten all the knights and then perished themselves? Or the two populations might have co-existed in some sort of long-term equilibrium, akin to Forbes's "balance of nature"?

The question had famously been addressed nearly two centuries earlier by the Rev. Thomas Malthus, whose independent wealth and comfortable lifestyle as an English country curate gave him the leisure to consider what might happen as the population of the world grew ever bigger. His Cambridge degree in mathematics helped him to come up with an answer. It was appropriately encapsulated in the initial letters of his three children's names: Henry, Emily, and Lucy. As the population outstripped the food supply, he decided, the world would descend into the Hell of starvation.

Malthus presented his arguments and conclusions in his *Essay on the Principle of Population*, published in 1798. Like any good mathematician, he started by declaring his assumptions. "I think I may fairly make two postulata," he said. "First, That food is necessary to the existence of man. Secondly, That the passion between the sexes is necessary and will remain nearly in its present state."

Then came the mathematics: "Population, when unchecked, increases in a geometrical ratio. Subsistence increases only in an arithmetical ratio. A slight acquaintance with numbers will shew the immensity of the first power in comparison of the second." He concluded "that the power of population is indefinitely greater than the power in the earth to produce subsistence for man."

Malthus's law describes what happens when a population increases by the same ratio from year to year or generation to generation. To Malthus and his followers, that law spelled disaster, because it meant that the increase in population between succeeding generations grows

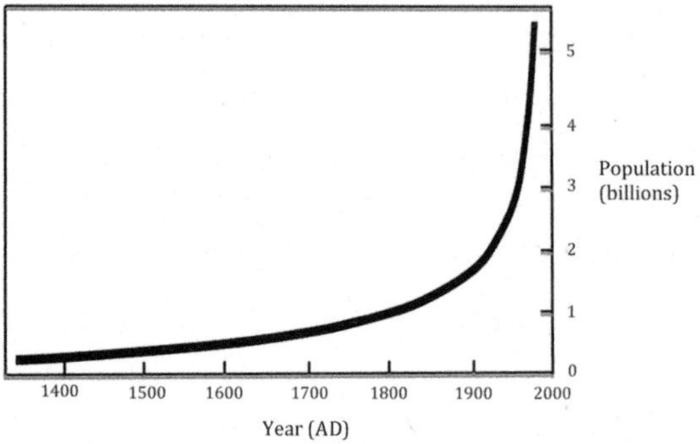

Figure 7.1. Human population growth curve over the last six hundred years. Adapted from David Price, "Energy and Human Evolution," *Population and Environment: A Journal of Interdisciplinary Studies* 16 (1995): 301–319, http://dieoff.org/page137.htm.

bigger with each generation and the growth curve becomes ever steeper, as seen in Figure 7.1.

This sort of curve is called an *exponential growth curve*. It describes a runaway positive feedback process where the rate of growth depends on the size of the population—the bigger the population the faster the growth.

Malthus argued that our food supply cannot increase at the same rate. Something has to give, and his depressing conclusion was that the population would grow until the rate of increase through reproduction was equaled by the rate of death through starvation.

Unless we limit our population voluntarily, argued Malthus, it will continue to increase until it begins to overshoot a sustainable level. The forces of nature will then conspire to bring the population back down, with a force that will increase as the level of overshoot increases. This limiting process, mathematically (and almost literally) speaking, is a *negative feedback* process.

Population growth, then, is controlled by a combination of positive and negative feedback processes, whose combined effects can in principle be used to predict its future course and whether that course is taking us toward disaster.

Malthus's equation gave only the positive feedback term. The Belgian mathematician Pierre François Verhulst added a negative feedback term* to Malthus's equation in 1838 and came up with an equation that now rejoices in the name of the *logistic difference equation.*

The equation can be expressed very simply in algebraic terms if we refer everything to a *maximum sustainable population* $\boldsymbol{p}_{\mathrm{max}}$. The population will continue to increase generation upon generation until it reaches that point. If we call $\boldsymbol{p}_{\mathrm{n}}$ the population reached at the nth generation, and $\boldsymbol{p}_{\mathrm{n}+1}$ the population in the following $(\mathrm{n}+1)^{\mathrm{th}}$ generation, then Verhulst's equation is:

$$(\boldsymbol{p}_{\mathrm{n}+1}/\boldsymbol{p}_{\mathrm{max}}) = r \times (\boldsymbol{p}_{\mathrm{n}}/\boldsymbol{p}_{\mathrm{max}}) \times (1-\boldsymbol{p}_{\mathrm{n}}/\boldsymbol{p}_{\mathrm{max}})$$

where r is the rate of population growth (if the population doubles each generation, for example, then $r = 2$).

We can see straightaway that, at low populations, where $\boldsymbol{p}_{\mathrm{n}}$ is small, the second term in parentheses (representing negative feedback) is approximately equal to one, and the equation reduces to:

$$(\boldsymbol{p}_{\mathrm{n}+1}/\boldsymbol{p}_{\mathrm{max}}) = r \times (\boldsymbol{p}_{\mathrm{n}}/\boldsymbol{p}_{\mathrm{max}}),$$

which is just Malthus's law for exponential growth expressed in algebraic terms. When the population becomes close to that which is maximally

---

* In mathematical terms, this makes the equation *nonlinear.* I have steered clear of using this term in the main text, but to scientists it represents the crux of the argument, since it means that the variables in the equation don't just change in simple proportion to each other. As we shall see later in the chapter, this means that very strange and non-intuitive things can sometimes happen.

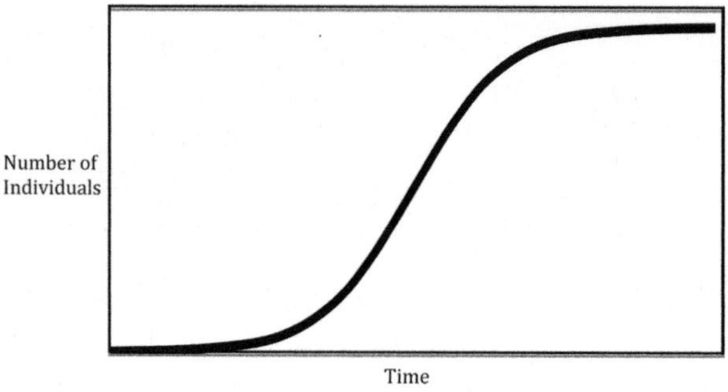

Figure 7.2. "Classical" sigmoid population growth curve.

sustainable, though, $(1-p_n/p_{max})$ gets closer and closer to zero, the rate of increase becomes slower and slower with each succeeding generation, and the population settles down to its maximal sustainable level. With the introduction of negative feedback, the exponential population growth curve has now changed to one that has a beautiful, simple, and above all understandable sigmoid shape (Figure 7.2).

ORGANIZED CHAOS

Th' whole worl's in a terrible state o' chassis.

—Captain Boyle in Sean O'Casey, *Juno and the Paycock*

The simplicity is deceptive, though, as May found out in the early 1970s when he started to plug some numbers into the equation. As he gradually increased the value of the growth rate r, the sigmoid shape steepened, reflecting the increasingly rapid initial rise in population. It became quite steep when the population doubled from generation to generation ($r=2$), and very steep when the population tripled with each succeeding generation. But when the growth rate r became greater than three, this

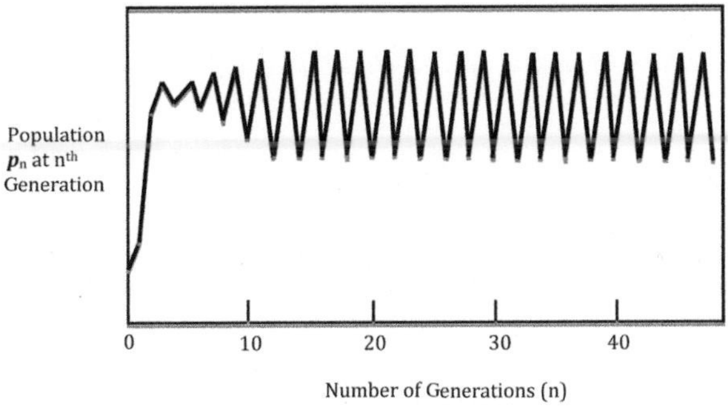

Figure 7.3. Oscillation of maximum population between successive generations for $r = 3.3$.

apparently simple equation started to give crazy, non-intuitive results. As can be seen in Figure 7.3, instead of *one* stable equilibrium population, there were now *two*, with the answer oscillating between them as time went on!

In other words, the equation predicted that the population would start going through a series of "boom-and-bust" cycles, such as are often seen in nature.

The traditional explanation for such cycles had been cast in terms of external factors, such as a change in food supply or habitat. Now, it seemed, some cases of boom-and-bust might be built into the very dynamics of population growth. A growing population might, by its very success, carry the seeds of its own demise. Luckily, it seemed from the mathematics that the demise usually carried the seeds of a subsequent recovery.

There was more to come. As May increased his hypothetical reproduction rate still further, he found that the population oscillated between *four* different values (Figure 7.4), then *eight*.

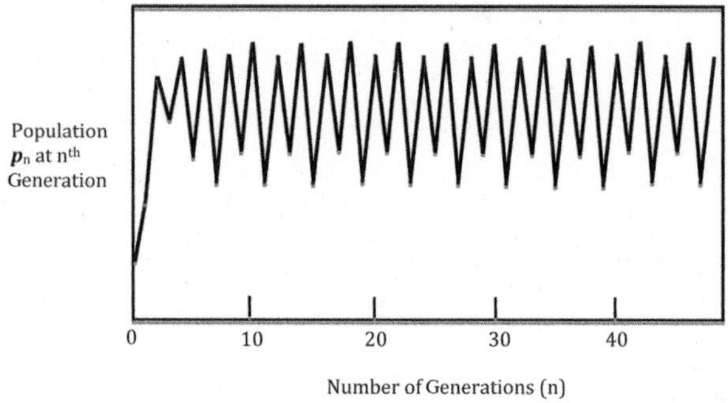

Figure 7.4. Oscillations in population for $r = 3.5$.

Eventually, at a critical growth rate of around 3.5699 (called the *point of accumulation*), the oscillations began to go hopelessly out of control, to the extent that May wrote a despairing message on a board in the corridor of the physics department: "What the Hell happens when lambda [a mathematical measure of stability that I have written as *r* in the equations above] gets bigger than the point of accumulation?" What had in fact happened was that he had discovered how *chaos* can emerge from a set of very simple rules—a conclusion that he hammered home to the scientific community in a seminal article entitled "Simple Mathematical Models with Very Complicated Dynamics."

The article stimulated a revolution, not least in providing one of the major mathematical foundations for chaos theory. May pointed out that the same sort of equation could be used to describe evolving relationships between commodity quantities and prices, the theory of business cycles, the spread of disease, and even the process of learning. All of these applications have since come to fruition to various degrees. The main initial impact, though, was in ecology, where the application of mathematics to the problems of ecology picked up a momentum that continues to increase.

Verhulst's population equation was only the beginning. It was just about the simplest nonlinear equation that one could visualize, with rules consisting of nothing more than one positive feedback process and one negative feedback process, working in tandem. May's results pointed up a problem that theoreticians are still coming to grips with: If these simple rules can produce such a complicated array of outcomes, depending sensitively on the initial conditions, what hope is there for predicting the outcome in real situations, where there are likely to be *many* feedback processes of both kinds going on simultaneously?

One thing was clear—writing down the appropriate feedback equations was just a start. Predicting the outcome of their interactions was the tricky bit. Fortunately, the age of computers had begun, and computer simulations have turned out to be an essential, and very useful, aid to prediction.

One of the first surprises from the early simulations was that natural ecosystems can have *many* possible stable states, rather than the one ideal state postulated by Forbes. The simulations also showed that a system can end up in any of these states; which one it ends up in seems largely to be a matter of chance.

John Sutherland from the Duke University Marine Laboratory at Beaufort Inlet, North Carolina, was one of the first scientists to confirm these predictions in the world of real ecosystems. The shoreline near his laboratory is home to a fouling community of hydroids, tunicates, bryozoans, and sponges that live on the intertidal rocks and also populate parts of the pirate Blackbeard's flagship *Queen Anne's Revenge*, resting in twenty feet of water off Fort Macon.

To study how such communities evolved, Sutherland devised a beautiful series of experiments. He simply hung a set of unglazed ceramic tiles about a foot under the water beneath the marine laboratory dock and watched to see what happened. Soon the larvae of different species found the plates, stuck to them, and began to grow. Some of the plates were initially colonized by tunicates. Others started off with sponges. All of the plates were eventually colonized by a mixture of species, which settled down in each case to a constant ratio as time went on.

The ratio was very different for the different plates, though, even though they were hanging in essentially the same environment. The final ratio was largely governed in each case by whatever larvae had happened to get there first. There was no convergence to a common ratio. If there is such a thing as "the balance of nature," then Sutherland's experiments showed that there are *many* balance points, not just the one.

Imagine a pinball machine, one with a bumpy base, full of hills and valleys. When a ball is fired in, it caroms off the pins (the "accidents of history"), rolls up and down over the hills, and eventually settles in one of the valleys (each of them corresponding to a different possible "equilibrium" state). If the machine is shaken slightly (a small "perturbation" from equilibrium), the ball may roll around, but it will remain trapped in its valley. If the machine is shaken vigorously (corresponding, for example, to a large-scale natural disaster) the ball will escape, roll around, and perhaps settle in a different valley.

This "ball in the cup" analogy should not be taken too literally, but it provides a helpful intuitive aid to understanding some of the major processes involved. In scientific terms, the valleys are called *stable states*. The maximum sustainable population predicted by Verhulst's equation (Figure 7.2) is one example of such a stable state.

Mathematicians call stable states *attractors*—a fairly obvious metaphor that reflects the fact that the system is drawn to them. Complex systems such as economies, societies, and ecosystems can have many alternative attractors. Critical transitions happen in conditions where the system can shift with ease from one attractor to another, thus switching between alternative stable states. These conditions are represented in our pinball analogy by the ball being either on top of a hill or in a saddle between two hills that is the equivalent of a mountain pass.

In the real world, the transition between one stable state and another can sometimes be accomplished with nothing more complicated than a mirror. When my daughter was young, for example, she had a pet peacock whose continuous calling was a permanent irritation, to the extent that we eventually gave the bird to a neighbor—ostensibly as a gift, but in reality as a reprisal for some now-forgotten offense. The ploy backfired when our neighbor provided the peacock with a mirror, effecting a transition from a permanently noisy state to an alternative silent state—the bird spent all day just looking at itself.

Mirrors can also be used in more serious circumstances. Captive flocks of flamingos are known not to breed unless their numbers and density are above a critical level. Many years ago, the flock at Taronga Park Zoo in Sydney was below this level and looked to be in danger of disappearing until a bright keeper had the idea of surrounding the flamingo cage with large mirrors to give the illusion of greater numbers. The birds fell for it and at last count were breeding happily—and no longer with any need for mirrors.

The situation faced by the flamingos is called the *Allee effect*: A population can grow if it is above a critical level, but will go into free fall if it goes below this level. The effect doesn't just apply to animals. It can apply to plants in a rain forest, to plants in harsh environments (where a critical density can be needed to "engineer" the microclimate), and even to scientists with their need for collaborative interactions.*

---

* The "new" Cavendish physics laboratory at Cambridge is designed to ameliorate the Allee effect through its arrangement of rooms and laboratories, which are configured so that scientists can't avoid bumping into each other whenever they come out of their laboratories or offices.

The two alternative stable states in the case of the Allee effect are those of growth or extinction. An equally extreme set of alternatives arises from the climatic *ice-albedo effect* on the temperature of the earth, which is largely governed by a balance of incoming radiation from the sun and heat loss back into space. This balance is affected by the coverage of ice and snow. With more ice and snow, more radiation is reflected back into space, and the earth's surface gets colder, promoting the formation of more ice and snow. The flip side is that, as ice and snow melts, the earth's surface gets darker, absorbing more radiation, getting hotter, and melting still more ice and snow. This is only one of the many processes that control our climate (see Chapter 10), but on its own it provides a good example of the "two possible stable states" scenarios that characterize the Allee effect.

The pinball analogy presents the Allee effect as a situation where the ball is on top of a hill and could roll into either alternative valley of stability with just a slight push. One thing missing from this simple metaphor is the recognition that the valley may not be a *simple* attractor (the ball eventually settles in the bottom of the valley), but a *strange* attractor (the ball keeps rolling around, following slightly different paths but never settling into one position). This image is more appropriate for many ecosystems, where "stability" consists of dynamic evolution through a series of closely related, but not identical, states.

Fortunately, it does not need such complicated imagery (and the complicated mathematics that goes with it!) to understand the basic principle for forecasting critical transitions: Know the location of real-life hilltops. One of my personal hilltops occurred when I met my present wife through a fortuitous set of circumstances after a conference in England where I was a visiting speaker from Australia. Our meeting set up two "alternative states"—to live in England or to live in Australia. Minor circumstances decided us for England, although we now lead a strangely attractive life in which we cycle between the two.

This story also illustrates one thing that is missing from the pinball analogy: The bases of the pinball machines, representing nature and society, are not static objects. Their profiles evolve with time, as do our lives, economies, and societies. What was once a deep valley may rise in time to become a hill. A ball that was trapped in the bottom of the valley now finds itself sitting on top of a hill. A slight movement may now be enough to send the ball rolling into any one of several nearby valleys. As the Allee effect demonstrates, some of these valleys may represent a comfortable equilibrium state, while others may represent disaster.

How can we predict when we are getting perilously close to such a transition? The answer lies in the use of *models*—simplified pictures that capture the essence of a situation and let us see through the complexities. What we can see, though, depends on the type of model we use. In the following chapters, I examine three main approaches—one that may help us to use our intuition in visualizing situations (*catastrophe theory*), one that can help predict the outcome of complex situations (*computer models*), and one that helps us to act like toads in perceiving and acting on warning signs in advance of disasters (*early-warning signals for critical transitions*). There is a fair bit of overlap between these approaches, but broadly speaking:

Catastrophe theory is the science of the pinball machine with the bumpy base. Its basis is heavily mathematical, but its usefulness outside the mathematical arena resides in the fact that it provides *images* that help us visualize the occurrence and consequences of critical transitions in an intuitively satisfactory way. In Chapter 8, I show how we can use these images to help understand love-hate relationships, binge drinking, and much more.

Computer models allow experts to make predictions in situations that range from personal relationships to economic forecasting or the future of the global ecosystem. Some of these models are based on catastrophe theory, but this is not by any means the only approach. In Chapter 9, I examine how models are constructed and suggest a set of rules that we can use to distinguish the brilliant from the bogus.

Early-warning signals for critical transitions have also come from computer modeling, but we don't need computers to use them. In Chapter 11, I outline the major signals that have been discovered and examine how we might apply them in everyday life. One day they may even help us to predict disasters almost as well as toads can.

8

Teetering on the Brink of Catastrophe

Anything in history or nature that can be described as changing steadily can be seen as heading toward catastrophe.

—Susan Sontag, *AIDS and Its Metaphors* (1990)

Catastrophe theory provides a way of classifying catastrophes so that we can visualize them pictorially, understand their different characteristics, and get an intuitive feeling for how they arise and what we might be able to do about them. Invented by the eccentric French mathematician René Thom in the early 1970s, the theory grabbed the public imagination in much the same way that Einstein's theory of relativity had done near the beginning of the century. Like the theory of relativity, catastrophe theory addressed a grand question and made a bold intellectual leap to reach a new and unsuspected answer. Like the theory of relativity, its central idea seemed to be implied by its attention-grabbing name. But also like the theory of relativity, those who took the name literally missed the point.

Many nonscientists still seem to think that the theory of relativity leads somehow to the conclusion that "everything is relative." In fact, there are two theories—the special theory and the general theory—and neither of them leads to this conclusion. The special theory tells us that the speed of light appears invariant to us, no matter how fast we are moving toward or away from the source. The general theory interprets gravity as a geometric property of space-time.

The conclusions from both theories of relativity have had a great impact on our view of the world, especially on the philosophical notion of simultaneity and, in more practical terms, on the equivalence of mass and energy that underpinned the development of the atomic bomb. The sort of catastrophe that might ensue from a nuclear war, however, is not the principal focus of "catastrophe theory."

Catastrophe theory is not just about catastrophes. In the terminology of this book, it is about critical transitions, which include real-life disasters and catastrophes but are not limited to them. Thom's theory lets us draw critical transitions as pictures—pictures that anyone can understand even if they can't understand the deep mathematics that lies behind them. Fortunately, no one needs to be a mathematician to understand the meaning of the pictures.

Their mathematical justification is deeply buried in Thom's book *Structural Stability and Morphogenesis.* I bought my copy of the English translation as soon as it became available in Australia in 1975. Catastrophe theory seemed to be at the cutting edge of science. I wanted to be right there with it, especially since Thom grandly announced in his foreword that his book was the intellectual successor to D'Arcy Thompson's *On Growth and Form*, written in 1917. Thompson's book was still a primary reference for those of us who were keen to unravel the way that the shape and function of biological organisms were intertwined. I hoped that Thom's book would help to provide an answer.

When I looked in its pages, however, I often could not make heads or tails of what I was reading. I found powerful and deep mathematics, but the relevance to biology seemed obscure.

I was not the only one who struggled to find the relevance. Later I found that the Nobel Prize winner Francis Crick, famous for his use of mathematics to help unravel the structure of DNA, had had the same problem when he met Thom. "My impression of René Thom," Crick wrote in his 1988 autobiography, "was of a good mathematician but a somewhat arrogant one, who disliked having to explain his ideas in terms nonmathematicians could understand. . . . He seemed to me to have strong biological intuitions but unfortunately of negative sign. I suspect that any biological idea he had would probably be wrong."

Fortunately, the English mathematician Christopher Zeeman was on hand to help clear the fog and explain the true significance of catastrophe theory. The "explanations" about biological growth and form were largely a fancy of Thom's, who was determined to show how important his new ideas would be to biology. Zeeman highlighted the real importance of Thom's book, where Thom showed that *all* catastrophes (up to a particular, quite high, level of complexity) can be classified into just seven "elementary" types, each of which can be represented in graphical form. There are the fold, the cusp, the butterfly, and the swallowtail, whose evocative names reflect their shapes. (Salvador Dali's last painting, *The Swallow's Tail*, was based on the swallowtail catastrophe.) There are also the hyperbolic umbilic, the elliptic umbilic, and the parabolic umbilic, whose names are not so evocative, except to mathematicians.

All of the types have elements of beauty that give them an instant aesthetic appeal. For the fold and the cusp, however, there is something more—a connection with real life. The cusp seems to be everywhere, and can even be seen on the surface of liquid in a saucepan or coffee in a coffee cup (Figure 8.1), while both the fold and the cusp have an intimate connection with the prediction of disasters.

RETURNING TO THE FOLD

In a "fold" catastrophe there are two alternative stable states, but no easy path from one to the other except near a critical transition. One of

Figure 8.1. Cusp on the surface of milk in a saucepan. Photograph by Len Fisher.

this type's most inimical manifestations is the *poverty trap*, where social and economic processes conspire to produce the scenario illustrated in Figure 8.2. It is a simplified picture, but it still encapsulates a lot of what happens in real life.

The origin of the word "fold" is easy to understand from the shape of the graph. This is not a conventional graph, though. The solid lines labeled "Wealth" and "Poverty" represent *alternative stable states*. Like the valleys in the pinball-machine analogy, they are *attractors*, and they epitomize the very real situations where great wealth and dire poverty can exist side by side, as they do in so many countries.

The role of the solid lines as attractors is shown in Figure 8.3: One way to think about this diagram is to imagine that you are looking down into a deep valley from a helicopter. The solid lines are rivers in the bot-

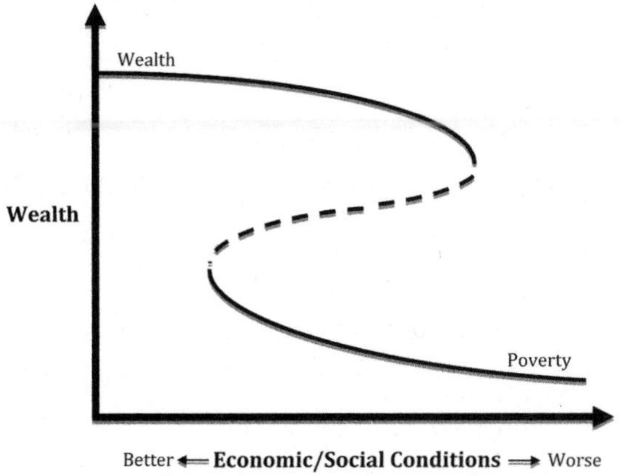

Figure 8.2. The poverty trap as a pair of fold catastrophes.

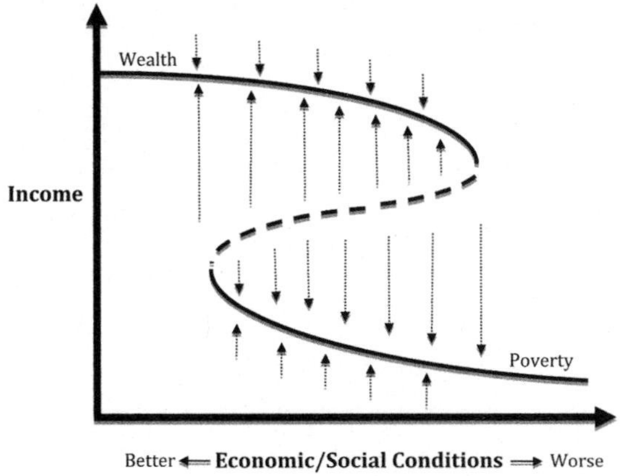

Figure 8.3. Wealth and poverty "attractors."

tom of a valley, while the dotted arrows are the direction of the flow of water down the hills on either side.

The dotted arrows illustrate the truth of the axiom "The rich get rich and the poor get poorer." If you have sufficient income, starting capital, or social advantage, then normal economic and social processes will draw you toward the "Wealth" line. If you don't have those advantages, you will be drawn inexorably toward the "Poverty" line.

The dividing line is the central, heavily dotted line, which might be a hilltop in our geographical landscape metaphor. It is a line of critical instability where just a small fluctuation in your fortunes can take you in either direction. A small increase in salary or a slight drop in interest rates may make your mortgage repayments easier and ensure your rise to wealth. A small increase in interest rates or loss of a job, however, may lead to your house being repossessed, and then you may end up in the poverty trap, where high rents prevent you from saving enough for another mortgage without help. Many people live perilously close to this line.

At the two ends of the dividing line lie critical transitions (Figure 8.4). A relatively wealthy person might lose his assets through gambling or the sudden collapse of the stock market (which amounts to the same thing) and end up at the right hand of the two critical transitions, crashing into a state of poverty. Someone with not much money might strike it lucky or work her way up (perhaps with the help of a buoyant economy) until she reaches the point where the left-hand critical transition takes over and positive feedback processes such as those involved in the spiraling growth of a business take over.

There is another way to shift from one alternative stable state to the other—one that does not require waiting for a critical transition. It is called *forcing* (Figure 8.5).

"Forcing" in the upward direction is a matter of providing a small injection of capital, education, or social advantage that is sufficient to take an individual or a nation from one side of the line of critical instability to the other. It is epitomized in the work of the Nobel Prize winner Muhammad Yunus, the Bangladeshi economist who came up with the idea of *microcredit*. Microcredit is the extension of small loans to poor people so that they will have enough capital to begin their own small

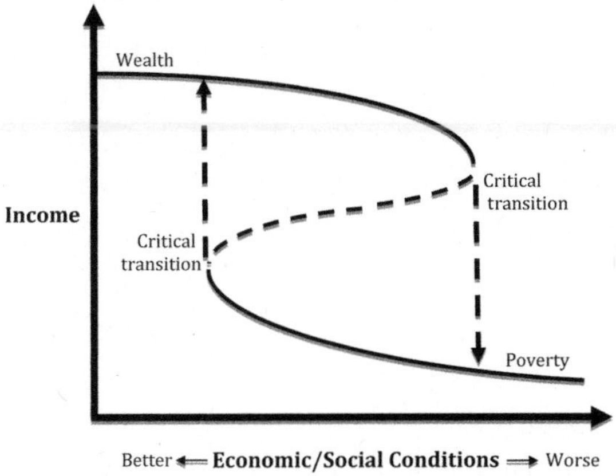

Figure 8.4. Critical transitions between wealth and poverty.

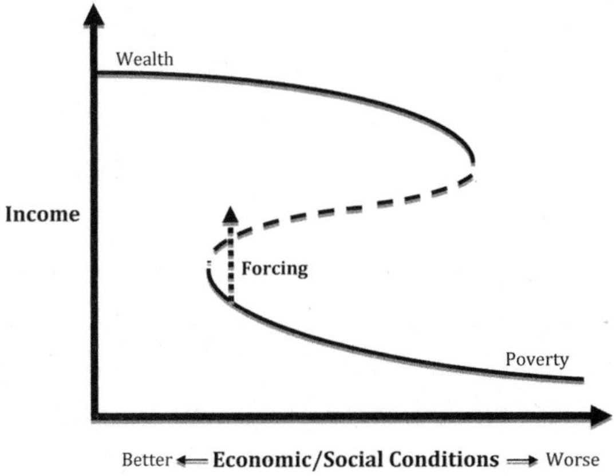

Figure 8.5. "Forcing."

businesses. It has been an enormously successful enterprise, and rather similar ideas are now being applied by economists involved in the restoration of damaged national economies.

This is not to say that catastrophe theory has all the answers. In most cases the theory is best seen as a guide—a model that lets us see the forest for the trees, even though, to make realistic progress, we must eventually consider the role of the individual trees. In this capacity, the fold catastrophe in particular has proved to be a very fruitful starting point. Ecologists use it to help understand how ecosystems can suddenly swing from one state to a very different state, as when a clear shallow lake such as the one studied by Forbes suddenly becomes dark and turbid. Psychologists use it to help understand mood swings, and it has been invoked to explain irrationally rigid patterns of behavior and even the collapse of ancient civilizations.

Adding a third dimension gives us further insights—as a friend of mine discovered after a skiing accident in the aptly named Perisher Valley snowfields, high in the Australian Alps. Jan was skiing back down to the lodge after a long day when a mist suddenly descended. Undeterred, he kept on skiing down the hill, confident in his ability to follow the trail. His confidence took a downturn when he found that he was traveling rather faster than he had anticipated, but that was nothing to his dismay when he found himself skiing on air instead of snow.

Luckily, Jan fell no more than ten feet before he hit snow again in a tangle of twisted skis and poles. Even more luckily, nothing was broken or dented except for his pride. When the mist lifted, he found that he had skied straight off an overhanging ledge with a cross-sectional shape that was very similar to the shape of a fold catastrophe (Figure 8.6, top).

As with a fold catastrophe, there was no direct return path, but when Jan looked to his right he found that the overhang gradually diminished and eventually disappeared (Figure 8.6, bottom), so there was in fact a route back. This sort of configuration is very similar to that involved in a cusp catastrophe, which describes the process when there is a second factor that can influence the height and depth of the fold.

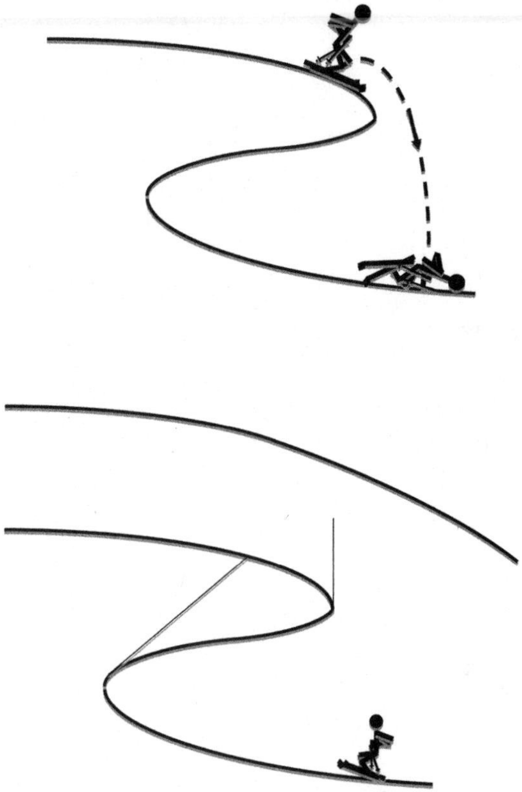

Figure 8.6. Skiing catastrophe: fold (top), cusp (bottom).

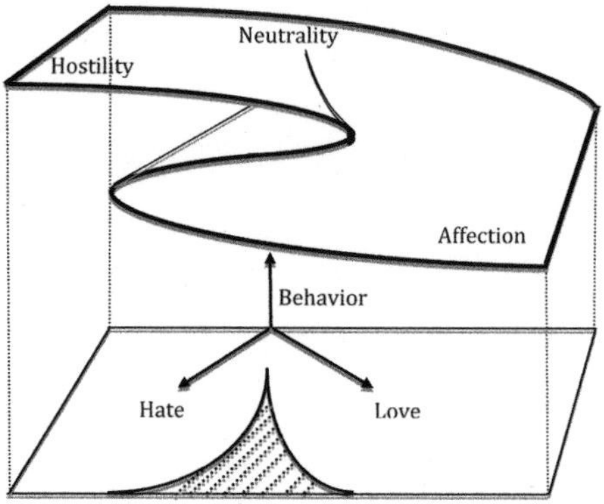

Figure 8.7. Love-hate relationship modeled as a cusp catastrophe. Adapted from Robert Galatzer-Levy, "Qualitative Change from Quantitative Change: Mathematical Catastrophe Theory in Relation to Psychoanalysis," *Journal of the American Psychoanalytical Association* 26 (1978): 921–935.

My favorite illustration is the use by the psychiatrist Robert Galatzer-Levy of the cusp catastrophe to portray what happens in love-hate relationships (Figure 8.7).

The graph is not as complicated as it looks! It is certainly not as complicated as real love-hate relationships, because it reflects only the basic issue of how our *behavior* toward someone can be influenced by a conflict between the emotions of *love* and *hate.*

Those three words—love, hate, and behavior—are drawn as three mutually perpendicular axes on the graph (like the three mutually perpendicular edges at the corner of a cube). These three axes (the lines with arrows) are sketched in 3-D fashion at the rear of the figure.

Now let's place ourselves somewhere in the love-hate plane (the lower quadrilateral) and draw a vertical line (in the direction of the behavior arrow) to indicate our response to the other person. We can make up our own conventions about what the length of the vertical line means. The convention here is that shorter lines represent affection, very long lines represent hostility, and lines of intermediate length represent neutrality.

As we move around the love-hate plane (technically called the *control surface*), the length of our vertical behavior line varies. Way over on the left, where hate predominates, so does our hostile behavior, and the vertical line is high. Over on the right, where love predominates, so does our affectionate behavior, and the vertical line is shorter. The varying height of the vertical line sketches out a behavior surface whose shape looks rather like that of the snowfield in which Jan found himself.

A fold is found in the middle, where there is an almost equal mixture of love and hate; here the behavior line passes through *three* points on the behavior surface. One represents affection, one represents hostility, and the middle one represents the unlikely possibility that both behaviors are on display at once—the equivalent to the dotted "line of instability" in Figures 8.2 to 8.5, where things can go either way.

With a fold, there come critical transitions. Just looking at the upper surface from the front (where both love and hate are at high intensity), we can see the obvious shape of the fold. If we move from love toward hate, we will reach a point where the curve folds back on itself, and our affectionate behavior will suddenly and dramatically switch to hostility via a critical transition. Once we are on the upper hostile surface, we have to move back a good long way toward love before the curve folds back on itself in the other direction and our affectionate behavior is restored.

Does this sound familiar? It does to me.

The third dimension offers an alternative, since it allows for both love and hate to be present at different intensities. In fact, if both become low enough, the fold disappears. We may visualize this by imagining a light shining from above, with the folded region casting a cusp-shaped shadow on the control plane, as shown in Figure 8.7. Once we are above the point of the cusp, the critical transitions disappear. Just calming down can help us to escape from the shadow.

The cusp catastrophe is admittedly trickier to follow than the simpler fold catastrophe, but it is a valuable aid to understanding and predicting the circumstances under which behavior can suddenly change when two strong conflicting emotions are involved. One of the earliest examples was based on the ethologist Konrad Lorenz's observations of the behavior of dogs when they are confronted with situations that induce both fear and aggression—they can switch without warning from submission to attack via a critical transition. Terence Oliva and Alvin Burns have also used cusp catastrophe theory to describe our rather similar behavior as consumers when we have a choice between complaining or meekly accepting a situation.

The psychologists Kelly Smerz and Stephen Guastello have used the cusp catastrophe to help understand the conflicting emotions involved in binge drinking among college students (Figure 8.8):

This graph isn't just a vague picture, drawn in intuitive fashion (although it is a considerable aid to intuition), but one that is based on a thorough statistical study. It paints a very different picture from that which psychologists have associated with binge drinking in the past.

Earlier models made the naive assumption that influential factors like attitudes to drinking and responsiveness to peer pressure could simply be "added up" to predict the likelihood of binge drinking. Mathematicians call this sort of model *linear*—that is, if we double the stimulus, we double the strength of the response.

The picture in Figure 8.8, derived from the real behavior of students, paints a very different, and highly *non*-linear, picture. Here, for example, we see that at high levels of peer pressure, a slow reduction in disap-

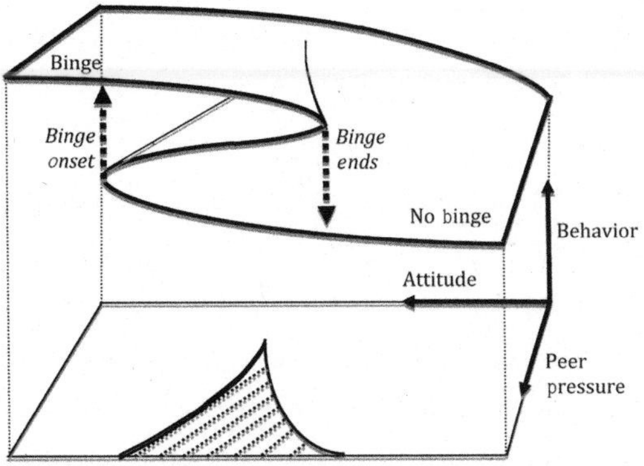

Figure 8.8. Binge drinking modeled as a cusp catastrophe. Adapted from Kelly E. Smerz and Stephen J. Guastello, "Cusp Catastrophe Model for Binge Drinking in a College Student Population," *Nonlinear Dynamics, Psychology, and Life Sciences* 12 (2008): 205–224.

proval of binge drinking (moving from right to left along the bottom axis) has little effect on the chance of binge drinking until a critical level is reached, and then the chance of binge drinking suddenly jumps. Just *where* the jump happens also depends on how much peer pressure is being exerted. If the peer pressure is low, there is no jump, and the transition is smooth.

Another difference from the mathematically (and psychologically!) naive linear picture is that the effect of attitude on binge drinking is not reversible (mathematicians call this *hysteresis*). Once binge drinking behavior starts, it takes a big change in attitude to stop it.

All of this may seem intuitively obvious. Perhaps it is. But the very fact that earlier psychologists accepted a naive linear picture of binge drinking behavior shows that a picture reflecting the more complicated reality has its value, especially if that picture is to be used as a guide to policy.

Catastrophe theory provides us with images that help us understand what is going on in situations that involve critical transitions. In the early days after the theory made its appearance, some people became overenthusiastic about its potential to do even more, and for a while its reputation was tarnished.

Now there is a clearer vision. Under appropriate circumstances, the theory can be used to derive quantitative information, as ecologists are now doing for critical transitions in relatively simple ecosystems such as shallow lakes. Even when quantitative information cannot be derived in this way, the images provided by catastrophe theory can still provide insights that our intuition may not lead us to suspect—so long as those images are used with circumspection.

When it comes to making policies, or producing arguments for changing policy so as to anticipate and avoid disasters, we generally need more concrete information. This is when we turn to mathematicians and their *computer simulations.* These may or may not be based on catastrophe theory, as the examples in the following chapters show.

9

Models and Supermodels

Some say the world will end in fire,
Some say in ice.
From what I've tasted of desire
I hold with those who favor fire.
But if it had to perish twice,
I think I know enough of hate
To say that for destruction ice
Is also great
And would suffice.

—Robert Frost, "Fire and Ice"

Robert Frost's view of the world's future strikes a powerful chord, especially for those of us who are worried about what the future holds for us and our children. According to many computer models, we are already heading for the "fire" of global warming. In the longer term, some models show, there is also a possibility that we could eventually return to the icy nightmare of a "snowball earth."

What are these models? How do they work? How can we judge them? Which of them are scientists' flights of fancy, and which should we take seriously and act upon? Part of my reason for writing this book was to find good answers to these questions, so that we can make better-informed judgments about these critical issues and about many other issues where computer models are used to predict abrupt and sometimes disastrous change.

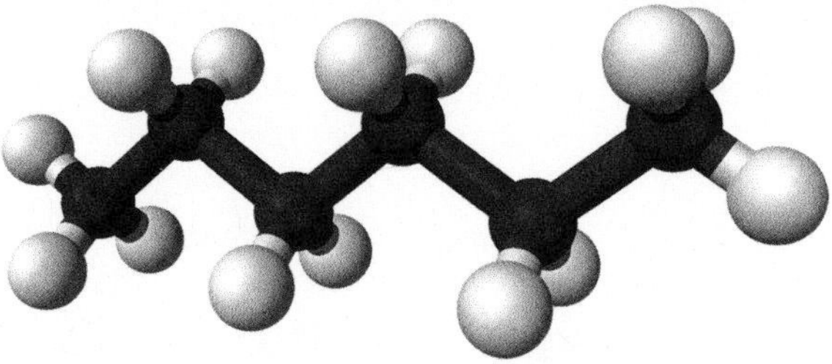

Figure 9.1. A "ball-and-stick" model of a hexane molecule: six carbon atoms (black) are linked in a line, with hydrogen atoms (white) linked to each carbon atom.

Models are metaphors for reality. They let us pick out the important aspects of a problem and bring them to the fore. In doing so, they can give us a new insight into the problem, just as literary metaphors do.

PHYSICAL MODELS

The simplest models provide a physical image that is an analogy for the real thing, such as the "ball-and-stick" models that chemists like myself use to work out the three-dimensional shapes of molecules (Figure 9.1). I have sometimes dropped off to sleep dreaming of such models, but never with the success that the German chemist August Kekulé had when he dozed off in front of a fire while trying to work out how the six carbon atoms in the benzene molecule are joined together.

According to his own account, his mind's eye "distinguished larger figures in manifold shapes. Long rows, frequently linked more densely; everything in motion, winding and turning like snakes. And lo, what

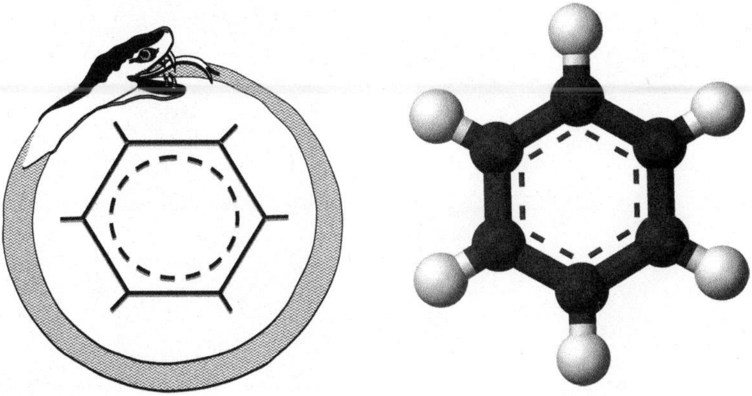

Figure 9.2. Kekulé's dream (left). "Ball-and-stick" model of a benzene molecule. Courtesy of Ben Mills (right).

was that? One of the snakes grabbed its own tail and the image whirled mockingly before my eyes. I came to my senses as though struck by lightning."

Kekulé came to his senses because the snake image let him see that the six carbon atoms could be linked in a ring, "tail-in-mouth" fashion (Figure 9.2). No one had dreamed of this possibility before, and the idea proved crucial in understanding the chemistry of this important molecule.

Of course, benzene is not made of snakes. They merely provided the image, just as cardboard cutouts did during the search for the DNA structure by Jim Watson and Francis Crick.

Their problem was to find out how four flat molecules called "bases" could fit together inside the double-helical structure. Crystallographers sought the answer by using X-rays to probe the structure of the actual molecule. Watson literally took a shortcut by cutting out scale cardboard models of the four bases, then sliding them around on his desktop until he could get them to fit neatly together (Figure 9.3). This is an approach

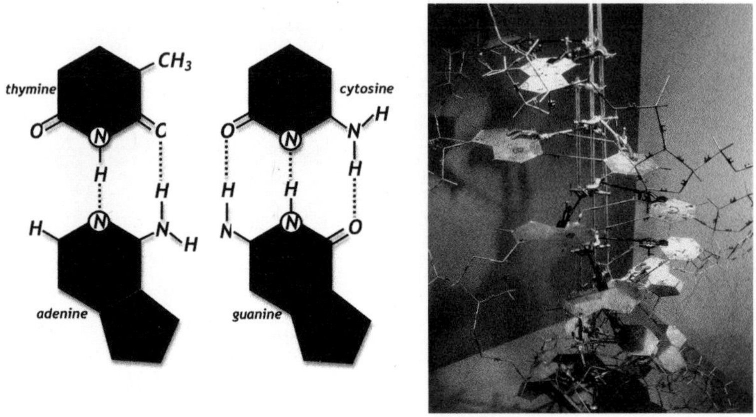

Figure 9.3. Simplified images of the four DNA bases (adenine, cytosine, guanine, and thymine), with some detail omitted to emphasize the molecular shapes and how they fit together in two pairs (adenine-cytosine and guanine-thymine), with each linked pair (linked by the dotted "hydrogen" bonds) having the same overall shape. This was the key to how the four differently shaped bases could fit inside the double helix—and also to the genetic code itself (left). Reconstruction of Watson and Crick's original scale model of the DNA helix, with the bases as flat metal plates. Science Museum, London (right).

that modern-day drug designers also exploit, although they use computer programs rather than cardboard.

On a grander scale, the geologist Alfred Wegener performed a rather similar feat to Watson's when he noticed that the continents on the earth's surface could be moved around to fit together like the pieces of a jigsaw puzzle. This led to the discovery of *continental drift*—a discovery that was as important in its way as the discovery of the structure of DNA.

Watson's pithy justification for his use of a simple model was that "nobody ever got anywhere by seeking out messes." He might have added that one of the most important aspects of simple models is that

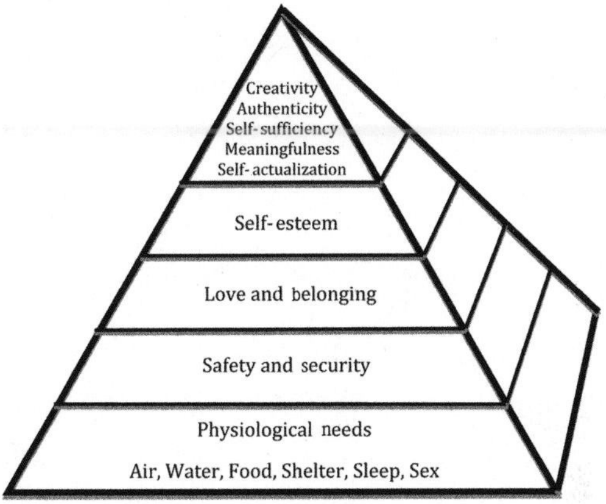

Figure 9.4. Maslow's hypothetical hierarchy of human needs.

they allow fruitful *predictions* that can be tested.* His and Crick's model predicted that the bases could stack as pairs of flat plates on the inside of the helix—a prediction that was confirmed by the building of another scale model, this time around six feet tall, and made of metal (Figure 9.3, right).

The use of physical images as models is not confined to the physical sciences. Psychologists also find them useful, as in the "pyramid" model introduced by the psychologist Abraham Maslow in 1943 to describe the hierarchy of human needs (Figure 9.4):

Maslow's model has been described as "one of psychology's genuinely good ideas." It has been criticized on grounds that include ethnocentricity, psychological realism, and the position given to sex,** and

* Many metaphors have been proposed for the structure of DNA, and DNA itself has now become a metaphor. There are books on the DNA of leadership, the DNA of innovation, and even the DNA of customer experience. All of these metaphors play on our image of DNA as the fundamental material that makes everything else happen. But the models that they espouse are not susceptible to test. They provide a shorthand description, and perhaps a useful encapsulation, but nothing more.

** Maslow's model predicts that, in a contest between sex and love, sex will take precedence. Later models by other authors avoid this controversial conclusion by putting sex above love.

Maslow himself later revised his model. The image that it provides still retains a powerful hold on the imagination. It is used by counselors and taught to marketing students in MBA courses, and it has even been invoked to explain the design of Chinese furniture.

As these examples show, physical models are a great aid to the intuition. Their metaphorical value can be misused, however, as happens with Maslow's model when it is invoked to "explain" the present concern of many people about the possibility of global warming. According to one skeptic, "The higher we move up the Maslow pyramid, the greater is our need for theoretical, manufactured, future 'catastrophes.' 'Global warming' thus comprised the classic 'need' of the 'loads-of-money' generation. But, we are now plunging back to reality, so that we no longer require this particular 'catastrophe' for our psychological fulfillment."

In this case, the metaphor is being used by a group of economists who seem to be primarily concerned with the effect on profits if policies to handle the threat of global warming are implemented. Maslow's metaphor could be interpreted another way, however. Global warming threatens our physiological needs, such as adequate food and shelter, and *this* is why it is so high on the agenda of many people.

MATHEMATICAL MODELS

When you can measure what you are speaking about and express it in numbers, you know something about it, but when you cannot measure it, when you cannot express it in number, your knowledge is of a meagre and unsatisfactory kind.

—William Thomson (Lord Kelvin)

The phrase "global warming" provides strong physical images of flooded cities, altered climate, and collapsed ecosystems. We can visualize how it *might* happen by using the metaphor of a greenhouse: Heat can get in, but it then remains trapped because of increasing concentrations of carbon dioxide, methane, and other "greenhouse gases" in the atmosphere. To understand whether this *will* happen, we need to make *quantitative* calculations—that is, we need to develop a *mathematical model.*

Mathematical models are usually based on physical images, but the connection can be quite tenuous. A model for the flow of a river, for example, may include equations that describe the change in flow rate with depth, width, and the presence of obstacles, all of which can readily be visualized. A model for the flow of money through an economy may include similar equations, but "width," "depth," and "obstacle" do not now have precise physical meanings. They are *metaphors.* "Obstacles," for example, may be laws that block the free flow of money, while "width" and "depth" may represent the number of people involved and the depth of their pockets.

Real economic models are much more complicated than this, of course. I made up this simple one to illustrate a point, which is that the development and use of a mathematical model usually proceeds in three stages.

The first stage is to develop an image of what is going on. This may be a physical image, or it may be a mathematical image. If it is a physical image (real or metaphorical), then this needs to be expressed as a set of equations. Events that we can visualize in terms of a fold catastrophe, for example—such as the fate of the skier in Figure 8.6—can be expressed mathematically as a set of equations that describe the mixture of positive and negative feedback processes that are controlling the skier's course.

The second stage is to solve the equations, usually numerically and with the aid of a computer. The equations are written as a software program, the starting conditions are entered—the skier at the top of the hill, for example—and then we push the button and watch to see what happens. In the case of the skier, the output provides a solution to the equations in terms of position and speed as a function of time.

The third stage is to interpret the output. This is easy in the case of the skier, but it can be much more difficult when we are trying to follow the evolution of an ecosystem or an economic system and want to determine how the course of that evolution changes when we vary the conditions. To predict when disasters are likely to occur, we need to identify scenarios that lead to critical transitions.

One such scenario occurs with buildings constructed in earthquake zones. When I made a radio program on the subject, I was interested to find that engineering students begin their training by constructing model buildings from dry sticks of uncooked spaghetti! The models are mounted on a "shaker" table that can be made to vibrate laterally, simulating the shearing stresses generated by an earthquake. One thing that students learn from these models is that a heavy weight at the top can help to prevent collapse because its inertia damps the lateral oscillations of the building. In real multi-story buildings, that weight is often a swimming pool, as in some top-class hotels in San Francisco and Tokyo.

A physical model helps to illustrate the principles and guide the design, but to analyze stresses quantitatively engineers need to use mathematical models. These are based on the use of Newton's laws of motion to calculate the forces and Hooke's law of elasticity to calculate deformations in response to those forces, with modifications and additions to account for the complex properties of real structural materials.

As Augustin-Louis Cauchy showed in the nineteenth century, the stresses and strains vary continuously throughout the structure. To calculate them, engineers use a process called *finite element analysis*. The structure is divided into small areas or volumes (the "finite elements"), and the forces and stresses in each segment are calculated as if the segments were separate entities. This is only an approximate result, because it does not account for the fact that nearby segments can affect each other. The results are put together, the near-neighbor effects are calculated, and these are then added in to provide the initial conditions for a new round of calculations. This process is repeated until all of the results for the different segments are self-consistent. Commercial software is now available to do the job for large structures.

Such detailed calculations have really only become possible with the advent of powerful computers; they are a far cry from the intuitive "seat of the pants" approach to design used by Gustav Eiffel, Benjamin Baker, and other nineteenth-century engineers. These calculations are relatively straightforward, however, compared to the problems that modelers face in predicting the outcomes of more complex situations.

One problem is the increase in the number of variables that need to be considered. In weather prediction, for example, the starting point is information from many weather stations about temperature, wind speeds, humidity, rainfall, and a host of other measurements. Even though the physical laws that control these processes are well known, the interaction of positive and negative feedbacks between all of them creates very complicated scenarios, with outcomes that are difficult to predict, especially near critical transitions. Even so, computer models of these scenarios are becoming increasingly realistic, and the overall accuracy of weather forecasts is improving at roughly one day per decade—that is, a forecast for, say, six days ahead is now as reliable as the five-day forecast was a decade ago.

When it comes to predictions that involve human behavior, the rules are much less clear, and computer models are correspondingly less able to provide accurate forecasts. Still, there *are* rules that can be used, such as those from game theory, which uses rigorous mathematics to predict outcomes when the proponents in a situation choose strategies that are guided solely by the logic of self-interest.

Some critics argue that game theory takes a narrow (not to say cynical) view of human motivations. "Predictioneer" Bruce Bueno de

Mesquita claims a 90 percent hit rate for predictions from his game theory–based computer models, which suggests that the cynicism is often justified. Other tests of game theory, however, have shown that people often act in a less rational, and sometimes more altruistic, manner and that the predictions of game theory are not all that reliable.

Weather forecasts are not all that reliable either. They may be improving, but forecasters still sometimes get it wrong. If they can get the weather forecast wrong, and if human interactions also enter the equation, how much trust should we place in the complex computer models that are used to derive predictions for the future of the planet?

The outcome of those predictions has been encapsulated by a large international group of scientists, who have put together a list of nine potential areas of calamity and used the latest computer models to take a hardheaded look at just how far we are along the road with regard to each of them. Some of their results are surprisingly reassuring. Some are alarming. And some should have us trembling in our boots.

The scientists have presented their report in terms of a "safe operating space for humanity" (Figure 9.5):

They point out that the earth has been in its present stable Holocene era for around ten thousand years—roughly the period over which civilization has evolved from practically nothing to its present state. During most of that time, the average temperature, freshwater availability, biodiversity, and other indicators of a stable environment have stayed within a narrow range. Now some of these indicators are starting to change rapidly as a result of human activities—in particular, our reliance on fossil fuels and industrialized forms of agriculture.

How will these changes affect our future? The scientists are looking at models of the earth's "resilience"—that is, its ability to recover from small changes, as it has done so often in the last few thousand years. It may be expected to continue to do so, unless we pass definite

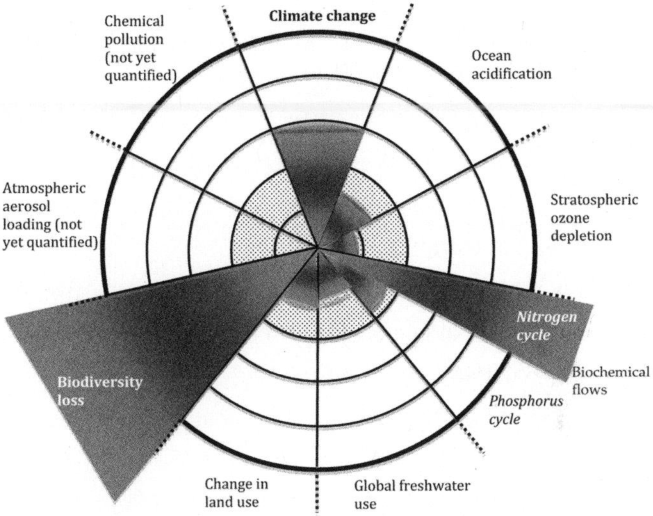

Figure 9.5. "A Safe Operating Space for Humanity," adapted from Johan Rockström et al., *Nature* 461 (2009): 472–475. The two central rings (shaded) represent the space within which it is safe to maneuver with respect to nine major environmental factors. The shaded wedges represent the actual state of affairs insofar as we are able to measure it. For three of those factors (climate change, the nitrogen cycle, and biodiversity), we have already passed the safe limit.

thresholds—labeled as "planetary boundaries" by the scientists—where runaway processes may take over to produce rapid and unwelcome change. The scientists who published this report have used the best information available to work out where we are now with respect to those boundaries.

There are some pleasant surprises. According to the models, we are still pretty safe with respect to stratospheric ozone depletion, global freshwater use, and even changes in land use. Where we are *not* safe is in climate change (we may already have passed a point of critical transition), the global nitrogen cycle (we seem to be well past the boundary), and biodiversity (we are in the throes of an outright disaster).

Does it matter? If we believe James Lovelock's *Gaia hypothesis* (which is itself a model), we can view the planet as a whole as a single gigantic organism with self-regulating processes that will maintain its internal balance against anything we can do. If Lovelock's hypothesis is true, one way in which the earth may maintain a balance is by ridding itself of disturbing irritants. If it treats us as one of those irritants, then yes, we should be worried.

As many of the examples in this book have shown, the shift from comfortable equilibrium or slow, steady change to abrupt, dramatic, and sometimes irreversible change can be upon us before we know it. If the "planetary boundaries" of Figure 9.5 are real, then it behooves us to take notice of them.

But how can we know whether to trust the computer calculations and the models on which they are based? How can we know whether the predictions are reasonable, or whether they may have arisen from mistaken assumptions or glitches in the mathematics? How can we know which prophecies are empty and which we should act on, especially when the predictions concern disasters that might befall our economies, our societies, and our planet? On what basis can we make an informed judgment when so much hangs upon that judgment and when skeptics are urging us to ignore the warning signs?

My answer as a scientist is that we need to use *applied skepticism.* In the following chapter, I suggest five skeptical tests that we can use to judge the validity of predictions based on mathematical computer models.

10

Beware of Mathematicians

Beware of mathematicians and all those who make empty prophecies.

—Saint Augustine of Hippo (AD 354–430)

Saint Augustine, the patron saint of brewers, printers, theologians, and sore eyes, was skeptical about the power of astrologers.* They were then known as *mathematici*, because they used mathematical methods to draw correlations and make predictions. Present-day computer-wielding *mathematici* use methods that sound suspiciously similar. Should we, then, follow Saint Augustine's example and be similarly skeptical about their predictions? Here I suggest four tests that we can apply to distinguish the gold from the dross:

1. Are the data reliable?
2. Is the model reliable?
3. Are the calculations reliable?
4. Are the people reliable?

Readers can judge the usefulness of these tests by applying them to the following story, told to me with relish by a Catholic priest.

* Saint Augustine was not always saintly. In his youth, he had a series of lovers, and at one stage he uttered the famous prayer: *Da mihi castitatem et continentiam, sed noli modo* (Grant me chastity and continence, but not yet).

Father O'Malley answers the phone.
"Hello, is this Father O'Malley?"
"It is!"
"This is the local tax office here. Can you help us?"
"I can."
"Do you know a Ted Houlihan?"
"I do."
"Is he a member of your congregation?"
"He is."
"Did he donate $10,000 to the church?"
"He will."

The predictive process and its outcome in this case pass all four tests. In this chapter, I ask whether more serious, computer-based models and predictions can pass the same tests. Inevitably there is some overlap between the tests, but I have done my best to separate the various strands.

TEST 1: ARE THE DATA RELIABLE?

If you can't trust your information, then you can't draw reliable conclusions from it. Failure to observe this simple principle has trapped many computer programmers and users of their products. It is encapsulated in "garbage in, garbage out"—a phrase usually credited to the early IBM programmer George Fuechsel. With an increasing belief in the infallibility of computers, the phrase has become transmogrified to "garbage in, gospel out," shortened to GIGO.

GIGO refers specifically to cases where impeccable logic leads to false conclusions because the logic was applied to false or unreliable data. One spectacular instance occurred on June 4, 1996, during the unexpectedly short maiden flight of the new Ariane 5 launcher. The rocket veered off its flight path about forty seconds after takeoff, broke up, and exploded—all because the input data for the flight path were being calculated by a computer routine that had been reused from the Ariane 4 vehicle, whose launch trajectory was different from that of Ariane 5.

It doesn't take a computer to foul up the input data. People are pretty good at it too—not just because of human error (which can be checked and corrected for), but also because of *selectivity.* When the Newfoundland cod fisheries closed in 1992, for example, fishermen and some scientists had been aware of the impending collapse for years. The public and the decision-makers did not see the full picture because there were incentives for the fishermen to underreport their bycatch, and institutional practices were also in place that selectively filtered the evidence.

Another well-known example concerns reports sent by troops in war zones, which are frequently over-optimistic and consciously or unconsciously designed to reflect well on the performance of the troops rather than convey the real situation. A surprising example also occurs when businesses pay good money for a computer model of their company's performance, but then provide the modeler with selective information so that the model's predictions will look good to board members and shareholders.

Even academic institutions are not immune. A group of angry unemployed lawyers, for example, have claimed that some law schools deliberately inflate predicted graduate incomes by including the incomes of students who get temporary jobs at relatively high pay rates but who can't find a permanent job.

Selectivity can also, however, be a plus. The political scientist Nolan McCarty has attributed the success of Bruce Bueno de Mesquita's game theory–based predictions to his careful selection of input data. De Mesquita does not claim to be able to forecast movements in stock markets, the outcome of elections, or the onset of financial crises where large numbers of people are involved. He mostly confines his predictions to situations with just a few opponents in different positions and consults widely with experts about the details of those situations. He is also an experienced and widely respected political scientist himself, with a good intuitive understanding of the negotiating strategies that the various proponents might use—strategies that he feeds into his models. This is not to decry his extraordinary results, although it may help to explain them.

The surprising outcome of recent studies on the science of prediction is that fewer data can lead to *more* accurate predictions, so long as the most significant data are selected. The point is that the inclusion of masses of largely irrelevant data can lead to a situation where we can't see the forest for the trees and we start to see trends where none exist. This principle has been shown to be true in economic forecasting, social forecasting, and even in weather forecasting.*

Under some circumstances, it is simply impossible to include even all of the *relevant* data. Pity the poor physicist trying to work out how a particular molecule in a liquid will move under the combined influence of the billions and trillions of other molecules that surround it. Even accounting for the influence of a few dozen nearby molecules, whose interactions are strong enough to make a difference, has to be dealt with by assigning an average "mean field" value for the interaction forces with the rest of the molecules.

Ecologists have a similar problem when it comes to forecasting. In complex situations, there are usually many significant variables that cannot be measured but that need to be estimated. Unfortunately, they can't just be averaged, as the physicist averages intermolecular forces. Instead, ecologists estimate the missing parameter values by fitting the predictions of the model to field data in a process called "tuning."

This would be all well and good if there were just one set of "best" guesses for the missing parameters. Unfortunately, there are usually many sets of guesses that could equally well fit the model. The forecaster is quite

* I describe many cases in *The Perfect Swarm.*

likely to pick the wrong set, which is all very well for "fitting," but very dangerous when it comes to forecasting, with GIGO being a likely result.

Luckily, there is an alternative. Instead of jacking up the complexity by adding more and more variables, ecologists can go minimalist by cutting their models down to bare essentials that require only the input of *known* parameters, not a mixture of the known with guesstimates. The work of the decision theorist Gerd Gigerenzer and others on minimalist models for decision-making has shown that such predictions are still likely to have validity, even if many of the less important parameters have been ignored.

The real test for trustworthiness in these situations is whether the most important data have actually been measured, or whether they have been estimated. If they have been estimated, was it on the basis of factual evidence, or have the data been "back-fitted" to the model, with all of the attendant dangers of that method?

Most of the models that have been developed to help predict our future global picture pass this test. Their input data are *independent* of the model rather than selected to fit it. This is a good start, but it is by no means the only feature that is needed for the predictions of a model to be trusted. A second feature is *reliability* or, at the least, *believability*.

TEST 2: IS THE MODEL RELIABLE? IS IT BELIEVABLE?

The Proof of the Pudding

The most reliable models are those that we can check by results. When measurements conform to prediction, we have good reason to believe that we are on the right track.

Newton's laws are the epitome. They provide a model for how material objects move and interact that has passed all tests and continues to pass them except at speeds approaching that of light, when Einstein's modifications become significant. At lower speeds, they provide such accuracy that NASA still uses them to predict the trajectories of spacecraft through the solar system, Einstein notwithstanding.

The early World3 model, developed by three MIT scientists and publicized in the 1972 book *The Limits to Growth*, has also passed such tests. It was the first integrated global model that linked the world economy with the environment. Its predictions attracted worldwide attention and formed a platform for the environmental and sustainability debates that have continued to this day. Those predictions covered eight key areas: global population, birth rates, death rates, food per capita, industrial output per capita, services per capita, nonrenewable resources, and persistent pollution.

The model was heavily criticized for its "doomsday" forecasts, even though the authors themselves specifically disavowed such overinterpretation. They recognized the model's limitations and said that it was predictive "only in the most limited sense of the word. [The graphs] are not exact predictions . . . at any particular year in the future. They are indications of the system's behavioural tendencies only." Even so, many of the model's predictions have now been shown to be pretty close to the mark.

The predictions were based on a simple model for the dynamics of growth that was not too dissimilar from the one used by Malthus, except that the equations in World3 used real data for measured trends and conditions. A big difference from Malthus's model was the recognition that the different elements in the model could affect each other, and these effects were included in the equations to generate realistic feedback loops—something that has been a feature of all subsequent models.

Despite its limitations, the model has been surprisingly successful. According to the Australian analyst Graham Turner, "Analysis shows that 30 years of historical data [from the time of the initial report] compare favourably with key features of a business-as-usual scenario [one of the three key scenarios examined by the report] . . . which results in collapse of the global system midway through the 21st century."

If that's not food for thought, I don't know what is. But World3 is not the only model pointing in this direction. Many later models, including updates of World3 by the original authors, have made similar predictions.

Independent Verification

The fact that other models point to the same conclusions provides a second test for the believability of a mathematical model. This test was sorely needed by mathematicians themselves when it came to the proof of the famous *four-color theorem*, which states that no more than four colors are needed to color in any map so that no two regions with a common boundary are the same color.

Numerous attempts had been made to prove the theorem, but without success. The famous recreational mathematics columnist Martin Gardner even claimed to have *dis*proved the theorem by producing a counterexample, which he published in his "Mathematical Games" column for *Scientific American* in April 1975. A college mathematics tutor who did not notice that the date of the magazine was April 1 triumphantly presented Gardner's "disproof" to his class as the latest important discovery in mathematics. The next day one of his students turned up with the map neatly colored in—in four colors.

Kenneth Appel and Wolfgang Haken from the University of Illinois produced the first real proof in 1976 by first reducing the problem to a large but finite set of possibilities and then using a complex computer program to check that each of these possibilities needed no more than four colors. When the news came out, local mathematics students reputedly celebrated in the streets, let off fireworks, and overturned cars. Other mathematicians, though, were not nearly so happy with such an inelegant proof that depended so heavily on computer number-crunching. It was virtually impossible to check whether there might be a mistake in the computer program except by using another, equally complex program that might also contain some subtle error.

The doubts have been allayed as other proofs have emerged that still depend on computer evaluation of the possibilities, but in different ways. There are now several such proofs, and mathematicians generally agree that, together, they provide convincing evidence that the theorem is true.

Following the Rules: The Human Factor

A model based on rules that are known to work provides a third reason for believing in it. In the "safe operating space for humanity," for example, models such as those for ocean acidification or stratospheric ozone depletion are based on physical processes that are now well understood.*

But can we say the same for the processes of human behavior that appear to lie at the heart of many global problems? Could there be a sociological equivalent of Newton's laws that would allow us to predict the future course of events for human society?

The science fiction writer Isaac Asimov invented one such set of laws in his "Foundation" series with the fictional science of *psychohistory*, defined as "that branch of mathematics which deals with the reactions of human conglomerates to fixed social and economic stimuli." Given our propensity for irrationality, not to mention social decision paradoxes such as the well-known "Prisoner's Dilemma," it is hard to see how our long-term behavior could *ever* be predicted in such a way, but that hasn't stopped some mathematicians from trying.

---

* The fact that ozone depletion appears now to be well within safe limits can at least partly be ascribed to the impact of *The Limits to Growth*, which stimulated research that resulted in the 1987 Montreal Protocol limiting the industrial production and use of fluorocarbons.

One approach is *social thermodynamics*—the quest for social parallels to physical quantities like temperature and entropy. Sometimes useful insights arise, but the results can also be frankly hilarious. It is not in the least surprising, for example, to find that, after pages and pages of convoluted mathematics, one author reached the conclusion that the social organization of his home country was essentially perfect.

At a more serious level, efforts are being made to understand how social structures emerge and function by treating them as *complex adaptive systems* (systems, broadly speaking, that show "emergent" behavior and in which the whole is greater than the sum of the parts). These efforts, which sometimes attempt to use insights from neuroscience and evolutionary psychology, are still at a fairly basic level and are not yet suitable for inclusion in serious predictive models.

That leaves us with two choices when it comes to incorporating human behavior into global models. One is to concentrate on the measurable *effects* of human behavior, such as productivity, pollution, land use, the release of greenhouse gases, and the like. This is what most models do, including the ongoing reports from the United Nations Intergovernmental Panel on Climate Change and the wider-ranging World3 and the "planetary boundaries" models.

The other choice is to borrow methods and insights from one discipline and apply them to another. The images provided by catastrophe theory, for example, have been used by ecologists to help produce realistic computer simulations of the interaction of many positive and negative feedback processes in the evolution of ecosystems. The fact that fundamentally similar processes (mathematically speaking) are going on in economies and societies suggests that economic and sociological forecasters could learn something from the models used by ecologists, and vice versa.

Ecology for Bankers

A major step toward this goal was taken in New York in May 2006 when the Federal Reserve Bank of New York teamed up with the various U.S. National Academies and the National Research Council to bring people from many disciplines together in a high-level meeting to "stimulate fresh thinking on systemic risk." The outcome was summed up in an article entitled "Ecology for Bankers," written by Robert May in collaboration with his fellow ecologists Simon Levin and George Sugihara.

It was a far cry from May's article on the ecology of dragons, although its motivation was the same—to get people thinking, this time about the idea that "there is common ground in analyzing financial systems and ecosystems, especially in the need to identify conditions that dispose a system to be knocked from seeming stability to another, less happy state."

The authors argue that catastrophic changes in a system ultimately derive from how it is organized. The changes emerge from feedback mechanisms within the system and from linkages that are latent and often unrecognized. The important thing is that the same principles apply no matter whether the system is an ecosystem, a banking system, a social system, an electrical grid, or the Internet.

The authors point out that

> [catastrophic] change may be initiated by some obvious external event, such as a war, but is more usually triggered by a seemingly minor happenstance or even an unsubstantial rumor. Once set in motion, however, such changes can become explosive and afterwards will typically exhibit some form of hysteresis, such that recovery is much slower than the collapse. In extreme cases, the change may be irreversible.

In other words, so long as there are complex interconnections, there is always a chance that some small and apparently insignificant event will set off a catastrophic chain of events. Thus have some ecosystems quickly perished—but others have survived for eons. Wherein lies the secret of survival? And can we apply those secrets to other complex systems?

What is needed, says the report of the 2006 meeting, is "robustness against perturbation." "Ecosystems are robust by virtue of their continued existence," say May and his co-authors. "Identifying structural attributes shared by these diverse systems that have survived rare systemic events, or have indeed been shaped by them, could provide clues about which characteristics of complex systems correlate with a high degree of robustness."

These comments led one writer to suggest that banks of the future should not be staffed with whiz-kid economists and ex-rocket scientists, but with ecologists.* In fact, what is needed is concerted interdisciplinary effort to understand the key structural factors that produce resilience and the boundaries beyond which that resilience can no longer be maintained.

Also required are political understanding of the importance of these matters and political will to implement what is needed. Part of my reason for writing this book is to help with the understanding and to point toward the material that those with the will are going to need. Within this space I can do no more than point—primarily toward the fact that, regardless of their detailed predictions, *all* of the major models for our global future have come to the same commonsense conclusions: *There*

* This is not to say that the individual models are perfect. Just to give one example, quite small areas of vegetation can initiate a "positive feedback" cascade that can turn adjacent dry and bare areas into wet and vegetated ones, which may become large enough to have a significant effect on the local climate. This effect has rarely if ever been included in global models.

Most ecological models, however, are still much more realistic and comprehensive than the sort of economic modeling criticized by the decision theorists Nassim Taleb and Spyros Makridakis for its assumption that the trends of the past will continue into the future. Ecological models do not do this, and they can in principle predict "black swans" that come in the form of critical transitions.

are indeed boundaries that we cannot pass without danger, and there are indeed things that we can do to maintain and increase economic, ecological, and social resilience within those boundaries.

But how can we calculate those boundaries and that resilience and know that the calculations are reliable?

TEST 3: ARE THE CALCULATIONS RELIABLE?

This is where the real difficulties start. In assessing the relative trustworthiness of different computer models, we inevitably come up against the fact that the complex calculations involved can be almost impossible to check and that some small change in the assumptions may lead to a large change in the conclusions.

The problem becomes worse when there is a dominant model in which most of the participants believe. The belief can sometimes be so strong that material that is inconsistent with it is filtered out. This is what happened with measurements of stratospheric ozone concentrations above the South Pole in the 1970s. The hole in the ozone layer remained undetected for years because computer programs that analyzed the data were instructed to reject measurements that did not accord with expectations derived from the model then in use.

Unfortunately, dominant models tend to go with dominant bureaucracies. This is not to say that the models are wrong, but as one group of prominent ecologists has pointed out, there is an ever-present danger that "the multiple filters required to establish scientific credibility and leadership create a hierarchy in which increasing dominance is exerted by nodes in which influence and citations concentrate. Such repeated demands for decision making from a few increase efficiency but also reduce the diversity of potential responses."

A second problem is the tendency of scientists to focus on the things that they can measure and calculate and to ignore the rest. It is a tendency that is well grounded in experience. The story of Robert Millikan's Nobel Prize–winning measurement of the charge on the electron, for example, is a story of knowing which results to accept and which to discard as being due to experimental artifact.

In being selective about data, though, we are sometimes in danger of throwing the baby out with the bathwater. As Isaac Asimov once pointed out, "The most exciting phrase to hear in science, the one that heralds the most discoveries, is not 'Eureka!' (I found it!) but 'That's funny.'"

The third problem is that we don't know what we don't know. We "routinely fail to ask the questions that would prepare us even for the anticipation of big, important change." But just because *we* don't know doesn't mean that *nobody* knows.

The ecologists suggest that the answer to all three problems, insofar as there is an answer, is *diversity* of thinking. They point out that this answer has often served ecologists well. To give just one instance, the information provided by illiterate village hunters and loggers proved crucial to understanding the precipitous decline in numbers of the Madagascan giant jumping rat, a species unique to the island.

The ecologists' answer is supported by studies of swarm intelligence, which show that groups can often do better than most of their individual members when it comes to making decisions—so long as individuals' contributions to those decisions are *independent* of each other. The opposite side of the coin is the problem of *groupthink*, where a dominant character (or, in this case, a dominant model) can distort thinking and block alternative solutions from sight.

Groupthink is a potential problem for large enterprises like the United Nations Intergovernmental Panel on Climate Change, which receives inputs from 194 countries and has a complex filtering process. The plus side is that scientists tend to be pretty independent and critical thinkers; if there is a divergence of opinion—as there was in the case of the melting rates of Himalayan glaciers—it is likely to come out into the open, to the benefit of all concerned. The subsequent investigation of the Himalayan glaciers controversy also led to the recommendation to restructure so as to reduce bureaucracy and increase transparency—reforms that, again, can only be for the good.

On balance, the concerns about dominant models, the focus on the measurable, and the lack of diversity have some basis in fact and need to be addressed seriously in the years to come if better global models are to be developed. These concerns are not sufficient, however, to overturn the major conclusions of the present models.

If the data are trustworthy, the model is believable, and the calculations are sufficiently well based, that leaves us with just two areas of uncertainty: the interpretation of the results, and our trust in the people who are providing the results and their interpretation.

TEST 4: ARE THE PEOPLE RELIABLE?

"To err is human. To really foul up—it takes a computer."

When experts disagree over computer-generated predictions, whom should we trust? Whose interpretation should we listen to?

According to one U.S. federal judge, we should listen to the computer itself, since computers have a "prima facie aura of reliability." That was in 1969, and things have come a long way since. Most people are now well aware that human frailty can enter the equation, on both the input and output side.

My personal answer to the question is: "I don't know." But I can make a few suggestions:

1. Try applying my first three tests. These can usually give a clue as to the trustworthiness of the process and the prediction.
2. Look at the alternatives. Does one make more sense than the others?
3. Look at the connections of the people who are making the statements. Do they have an ax to grind, or are their statements likely to be independent?
4. Look at the scientific credentials of the person making the statements. Is she talking about something in her own field? Then the statement can be given more weight, because the last thing a reputable scientist wants to do is to lose her reputation through making silly or exaggerated claims. Is the person very senior in his field? Again, then, he is worth listening to. This is not scientific snobbery. Someone who is a member of a national academy is likely to have gotten there on merit, and his opinions (in his own field of expertise!) are usually worthwhile.
5. Where were the opinions published? If they appeared in a peer-reviewed journal (especially a prestigious one like *Nature*, *Science*, or the *Proceedings of the U.S. National Academy of Science*), they should be given more weight. Again, this is not scientific snobbery. If people are prepared to have their opinions challenged by others who are at least equally knowledgeable, and if they pass that challenge, then their opinions probably have some merit.

On the other hand, opinions offered during interviews with journalists should be taken with a pinch of salt.

These suggestions may seem pretty obvious, but the obvious can sometimes be overlooked, especially when opinions, emotions, and the logic of naked self-interest are conflated.

One of the gags circulating when the original proof of the four-color theorem was published was: "A good mathematical proof is like a poem—this is a telephone directory!" Personally speaking, I would rather trust a telephone directory when it comes to getting facts right. When it comes to *combining* those facts to generate an informative overall picture, however, I would put my faith in the poet to provide a crystal-clear view of the otherwise unfathomable.

When it comes to visualizing the future, computers are the poets of modeling. They can produce unexpected insights in even the simplest of situations, such as the oscillations in population that emerge from the apparently simple equation for population growth. They can help us to predict the sometimes unexpected consequences of interactions between various parts of a complex system, such as the interactions between our global societies, ecosystems, and economies. But can they give us genuine forewarning when we are approaching a point of critical transition? In the next chapter, I track down the evidence that they might be able to do just that.

11

Weak Signals as Major Early-Warning Signs

Every calamity is a spur and valuable hint.

—Ralph Waldo Emerson, "Fate," *The Conduct of Life* (1860)

Emerson was philosophical about calamities and saw them mainly as hints as to how we might improve our performance in the future. Personally speaking, I would rather have the hint before the calamity. Given the choice between being philosophical in the aftermath of disaster or having some sort of forewarning that will help me to anticipate it, I will take the forewarning every time.

The trouble is that most forewarnings only come in the form of hints. Analysts call them *weak signals.* They can be difficult to detect and easy to misinterpret, as was the suspiciously long silence that eventually led to the following snatch of dialogue:

"Was it something I said?"
"No."
"Was it something I didn't say?"
"No."
"Was it something I should have said but didn't say until I eventually said it in the wrong way?"
"Maybe."
(Groan) "I knew it!"

Weak signals are also easy to discount because they are often mixed up with a lot of other information. It's like the situation at a party where you are trying to follow a conversation while many other noisy conversations are going on nearby. It is fatally easy to lose the thread and switch off or start listening to another conversation altogether.

That's not a good idea when it comes to sensing the weak signals that could provide forewarnings of disaster. We need to recognize their importance, find ways to identify them, and know what to look out for.

Psychologists have identified weak signals in the form of stressful life events that can provide forewarning of emotional disaster as their effects accumulate.* The original list, called the Social Readjustment Rating Scale, put bereavement, divorce, and imprisonment at the top, with marriage, marital reconciliation, and retirement close behind. Later lists varied the lower order somewhat, but there is general agreement that the cumulative effects of these events on health can be significant. There is no mention of book-writing in these lists, but from personal experience I would put it up there with imprisonment, followed by divorce and bereavement when the book is finished.

Information about these particular weak signals is now widely available, and they can be assessed by the individuals experiencing them. In other cases, however, the information about weak signals must come from a variety of sources for their cumulative significance to be identified. One of the more distressing examples is failure to detect child abuse: Neighbors, family members and friends, and different social agencies may have all had pieces of the puzzle, but no one was able to put the pieces together in time.

The cumulative importance of weak signals is significant in many arenas, from the development and maintenance of individual relation-

---

* Based on Hans Selye's original idea (see Notes, page 189) that stress can produce damaging, cumulative, physiological changes.

ships to the detection of incipient changes in the global environment. The fundamental principles for assessing and responding to their significance were set out in a seminal article published by the business strategist Igor Ansoff in 1975.

Ansoff was concerned by the tendency of managers to rely on "strong signals," such as net profits, unit costs, and the level of sales for different products. These signals permit short-term planning, but Ansoff argued that they are useless for anticipating events like the emergence of a new competing product, a change in the structure of the market, or even a change in the attitude of the workforce.

Ansoff maintained that managers need a "weak signal mentality" to cope with such contingencies: "Individuals responsible for identifying issues must begin to listen with their ears close to the ground for early warnings of threats and opportunities." If the quality of work from suppliers is getting worse, for example, or absenteeism is gradually increasing, these could be early warnings that the business is starting to go downhill. There may be other reasons for these changes, but their very existence should be a warning to managers to keep their eyes open.

Early warnings in business are especially important when it comes to events that business strategists label as "wild cards," which are more or less equivalent to Taleb's "black swans." These are events that are perceived to be very unlikely to occur but that would have a great impact on an organization if they do—such as the explosion of NASA's *Challenger* space shuttle in 1986, which derailed the whole space program.

The explosion was eventually traced to a faulty O-ring seal, but the deeper problem lay in the administrative structure of the organization, which was not adapted to detect early-warning signs of incipient problems. The danger signs, which included a tendency to brush unwelcome problems under the carpet, fit a pattern that has frequently been observed by crisis management experts: "Long before its actual occurrence a crisis sends off a repeated and persistent trail of early warning signals."

Ansoff and others have identified four principles for identifying and responding to such warnings. The first is to collect information and opinions from many different, independent sources about possible wild-card scenarios and then use this information to work out the most likely weak signals that would warn of such a scenario. The second principle is to train people in responsible positions to watch out for these signals rather than concentrating solely on the strong signals that permit short-term planning. The third is to set up mechanisms to bring these managers together to share their observations and discoveries. The fourth—and perhaps the most important—principle is to build a capacity into the organization *in advance* for improvisation that will allow sufficient change to deal with the scenario *if* it arises.*

Similar principles apply when it comes to anticipating the larger-scale problems we may face in our economies, our societies, and our ecosystems:

1. *Identify* the potential wild-card scenarios where possible.
2. *Learn* to recognize the warning signs for upcoming critical transitions, understanding that these signs can sometimes be ambiguous and open to interpretation.
3. *Share* information about warning signs that have been detected.
4. *Use* this information to prepare a plan that includes a built-in capacity for improvisation *if* the scenario should arise.

* To give a simple example, when a wildfire threatened my home in Australia, I was able to stay and fight it because I had prepared an escape route in advance to use if things threatened to get out of control. Firefighting in a business context works best when analogous foresight has been used to allow for changing plans with changing circumstances. As some experts have noted, it is nearly impossible to invent crisis management mechanisms while the crisis is taking place.

SKIRTING AROUND THE PROBLEM

The shorter the skirt, the higher the interest.

—Fisher's Law of Diminishing Hemlines

One interesting signal that has been proposed is based on what I call the Law of Diminishing Hemlines, which says that the economic prosperity of Western societies can be gauged by the length of women's skirts: The more prosperous the economy, the shorter the skirts. Known to economists under the more prosaic name of the *hemline index*, it has been given an unexpected twist by the social analyst John Casti, who claims that shorter skirts and other indicators of "social mood" can be used as early-warning signs of coming social and economic change.

The hemline index was proposed in 1926 by the economist George Taylor, who drew his conclusion after a careful study of fashionable skirt lengths during the previous fifty years. The hemline index has become something of an urban legend, but one of the few with a reasonable basis in fact. Skirt lengths dropped to the ground during the Depression of the 1930s, for example, and rose to an equally dramatic extent during the prosperous 1960s.

Taylor and others assumed that changing skirt lengths *reflect* economic circumstances. Casti turns that assumption on its head with his suggestion that "it is how a group or population sees the future that shapes events." In other words, according to Casti's claim, it is optimism (signified, for example, by rising hemlines) that produces prosperity, not the other way round.

It's an interesting, challenging idea. Unfortunately, even if it works for other indicators of "social mood," it doesn't work for hemlines. Recent studies have shown that skirt lengths *follow* changes in the economy, with a time lag of about three years, and that there is *no* correlation between the height of hemlines and subsequent economic circumstances.

Hemlines could still be a useful early-warning signal, however. According to the analyst Sandro Mendonça, the earliest hints of forthcoming change in the business environment come from a sense of "environmental turbulence." Ecologists have similarly identified "environmental turbulence" as an early-warning sign for the onset of critical transitions in nature. Two features of such turbulence are fluctuations between different states and an increasingly frequent appearance of extreme states. Translated into hemlines, this means that it is not the height of the hemline that matters so much as the fluctuation of fashions and the extremes between which they fluctuate.

There is a neat way to test this hypothesis. The Scandinavian designers Anders Mellbratt and Nils Wiberg have produced an "eco-friendly" dress whose length and glamorous appearance can be adjusted by sliding the hemline up and down on a set of tiny wires to suit the economic climate. If a set of sensors were attached to the wires, a record could be kept of hemline position over time.

Eventually, of course, the movement of the hemline could create a fold catastrophe in the material. As it happens, the fold catastrophe also provides the basis for the much more serious business of identifying early-warning signs for economic, social, and global disasters.

Fold catastrophes (Chapter 8) are simple models that have been shown to mimic many of the important features of real-life critical transitions. A close examination of the way in which such catastrophes progress toward their dénouement reveals that the key factor is *loss of resilience*—that is, loss of the capacity to absorb shocks and still function.* This capacity

* More technically, resilience is "the capacity of a system to absorb disturbance and reorganize while undergoing change so as to still retain essentially the same function, structure, identity and feedbacks." One way to understand resilience is to think about the elastic that holds your underwear up. When the elastic loses its resilience and fails to perform its function, it doesn't take much for your underwear to fall down.

is related to the capacity to adapt and reorganize in the face of changing circumstances. Loss of these capacities is an obvious warning sign of disaster to come.

Mathematicians have identified five key early-warning signs that can indicate a loss of resilience. In addition to (1) increasing occurrence of extreme states and (2) fluctuations between different states, there are also (3) *critical slowing down*, (4) *changes in spatial patterns*, and (5) *increasing skewness in the distribution of states*. These are all statistical measures that can require a great deal of data to analyze, and it can be difficult to distinguish true from false positives in that analysis. These measurements often provide us, however, with our only realistic hope of picking up warning signs in many vital arenas.

The evidence for their significance has largely emerged from studies on model and real ecosystems, but it is becoming increasingly clear that this portfolio of warning signs is equally applicable to many other socially significant critical-transition scenarios. To understand how these signals arise and how we might use the information, we need first to look at the processes that produce and control resilience.

RESILIENCE

Loss of resilience means that a system is slow to recover after a perturbation and susceptible to being pushed past a tipping point if another perturbation comes along before it has recovered from the previous one. Think of the plight of a boxer who has been hit with a dizzying blow and has not had time to recover before the next blow comes along. A sequence of such blows can spell the end of the fight.

Another example is coral reefs, which are normally kept free of seaweeds and other fleshy algae by fish and other grazers. This system maintains the natural resilience of coral reefs against disturbances such as hurricanes by keeping them open for recolonization by coral larvae. If the grazers are lost (through overfishing, for example), the resilience is lost, and the reef can become covered with fleshy algae after the next disturbance. This happened to many Jamaican reefs during the 1980s, when algal coverage rose from a few percent to over 90 percent following a sequence of events that began with Hurricane Allen. The reef was just recovering from the initial disturbance, assisted by grazing from the long-spined sea urchin *Diadema antillarum*, when a disease struck the sea urchins and the algae were able to take over in the absence of alternative grazers.

Computer modelers have a neat way to view the loss of resilience that can be the prelude to a critical transition. To them it is a matter of the changing shape of the valleys ("basins of attraction") that represent alternative stable states in our pinball-machine analogy (Chapter 7).

No matter how it is viewed, loss of resilience is a powerful indicator of potential disaster ahead. The concept was introduced to ecology by Crawford Stanley "Buzz" Holling in 1973, but it has much wider application than that. We can even apply it to our own bodies. If we are weakened by an infection or an injury, for example, our ability to recover from a secondary infection is reduced. Another example, which we can illustrate by a diagram analogous to that in Figure 11.1, is our ability to control our body temperature (homeothermy). We may represent this (Figure 11.2) by a one-dimensional "landscape" in which there are three basins of attraction—healthy, death from cold, or death from overheating. The healthy option is fortunately very deep, but not very wide in terms of acceptable temperatures. If we lose the ability to control our temperature, as represented by the dotted line in Figure 11.2, the valley becomes much less deep, and we can easily be tipped into one of the two alternative states.

When it comes to more complex scenarios, there can be many causes for loss of resilience. Some of the most complex are the social-ecological

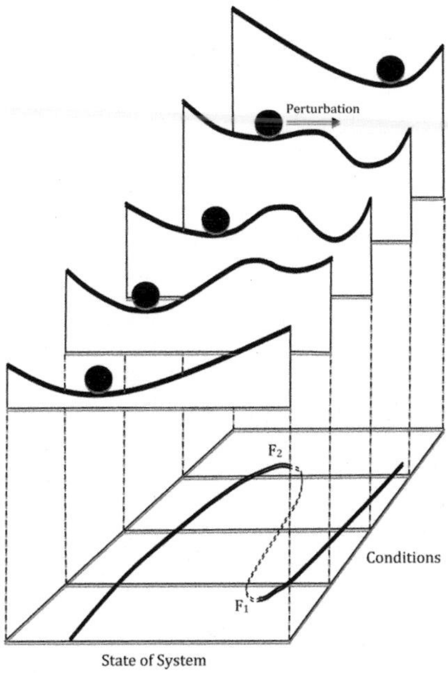

Figure 11.1. Changing resilience in the approach to a fold catastrophe. As the conditions change, the valley of stability becomes shallower and narrower, making it easier for the "pinball" to move to the alternative valley (representing the alternative stable state) with a small perturbation. Figure adapted from Marten Scheffer and Stephen R. Carpenter, "Catastrophic Regime Shifts in Ecosystems: Linking Theory to Observation," *TRENDS in Ecology and Evolution* 18 (2003): 648–656.

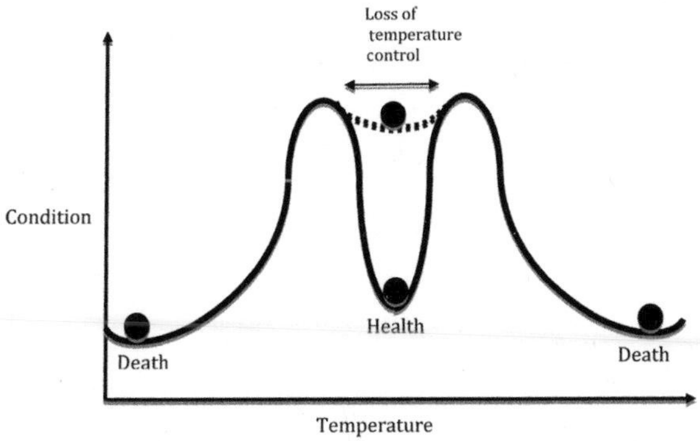

Figure 11.2. The valley of health and the valleys of death.

systems that constitute our living world. Here a loss of resilience may arise from cultural conservatism, specialization of species with regard to habitat, lack of diversity (as with monoculture crops), and a host of other interacting factors.*

These factors have been identified in a series of case studies by members of the Resilience Alliance—an international group of scientists studying the processes that produce or reduce resilience. One example of these complex interactions occurs in the Florida Everglades, where the wetland ecosystem has lost resilience through changes in the timing and volume of flooding. This has produced increasing concentrations of nutrients, a change in the vegetation, and a consequent decline in the numbers of nesting wading birds through loss of habitat.

On the other hand, there is high institutional diversity in the Everglades, with many organizations squabbling over water usage. This diversity has produced a "perversely resilient social system" that has been "stable and persistent" for the last forty years and promises to remain so in the foreseeable future.

Foreseeing that future is a complex problem, with resilience at its heart as a major key to sustainability. This is especially so with the biggest problem of all—the future of our planetary biosphere, which consists of all the living things on the planet.

The problem is being studied by international groups of scientists such as the Resilience Alliance and those continuing the work begun by the Millennium Ecosystem Assessment, which identified many "wild

* One of the more curious factors is the sacrificing of threatened animal species to help foretell the future. A particularly bizarre manifestation occurred during the 2010 soccer World Cup in South Africa, whose outcome was so successfully predicted by Paul the Octopus (Chapter 1). Some Africans who were gambling on the outcome attempted to improve their chances by killing endangered Cape vultures, putting their brains into cigarettes, and smoking them in the belief that this would help them to foresee the result—an action rather reminiscent of the oracular sniffing of volcanic gases in the cave at Delphi (Chapter 1).

cards" for environmental change and whose enlightening "Scenarios" reports may be found at http://www.MAweb.org. One of their main conclusions is that Stephen Forbes's "harmonious balance of conflicting interests" is insufficient in and of itself to maintain the resilience of an ecosystem (or an economic or social system) without the system being *managed* in some way to help sustain that resilience. Acting on early-warning signs is a critical feature for such management, and studies on model fisheries have shown that such action could indeed help us to turn back from the brink.

How that management can be achieved is another question. Top-down direction has failed in many more cases than it has succeeded, often because of the lack of flexibility that Ansoff and others have identified as a major cause of business failures. It has also failed when the significance of weak signals was not appreciated and acted on, as happened with the 2008 credit crunch: The increasing level of subprime mortgages provided a heavy hint about the upcoming crisis, although the significance of the hint was only appreciated in retrospect.

Consultative management has a better chance of success, since many more people, with different points of view, are involved in the search for the weak signals that can be a prelude to sudden and usually unwelcome change. As shown earlier, the most important and general of those signals have a distinctly statistical character, which identification from a variety of perspectives can help to resolve. Here I offer a brief account of some practical examples.

1. Increasing Fluctuations

Increasing fluctuations are driven by a process known as *stochastic forcing*: Fluctuations in the state of a system near a critical point can drive huge jumps between alternative stable states. It happens in one direction with the injection of microcredit; with "flickering" it can happen in *both*

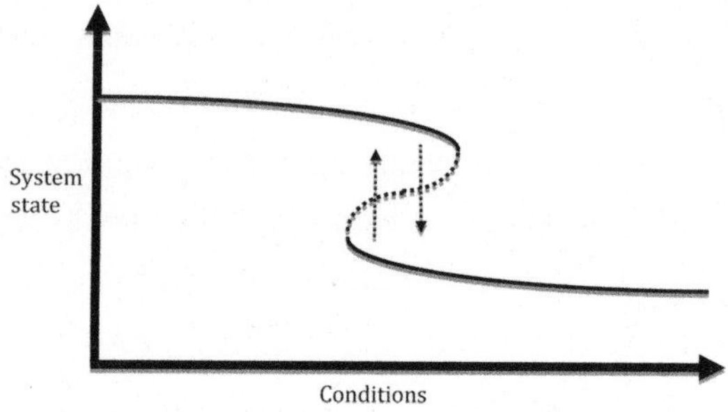

Figure 11.3. Flickering.

directions as fluctuations in the system state drive across the (dotted) line of instability (Figure 11.3).

The city of Manukau on the North Island of New Zealand bases its claim to fame on being the site for the country's major airport and having one of the oldest shopping malls in the country. Unbeknownst to most of its 385,000 residents, it has an even greater claim to fame: Manukau is the place where flickering was shown to be an unambiguous warning sign of a subsequent dramatic collapse in an animal population.

The animals were the polychaete tube worms, *Boccardia syrtis*, a species responsible for maintaining the integrity of the muddy sediments at the bottom of the harbor. The local ecologists Judi Hewitt and Simon Thrush had been monitoring the population for fifteen years at the behest of the Auckland Regional Council when the council stopped the low-level discharge from a nearby sewage outlet in 2001.

The animal community immediately began to fluctuate wildly in response to this minor change in circumstances. Two years later, the population of tube worms suddenly collapsed. With the loss of the tube-mats, the sediment became destabilized and susceptible to being stirred up by passing ships and other agents. It remains destabilized to this day.

Another instance of flickering is the fluctuation of fish populations between high and low numbers just before the collapse of a fishery. Unfortunately, the high phase can lead to an unwarranted optimism on the part of fishermen and legislators alike. If the true significance of the flickering had been understood and acted upon in time, some fisheries that have now collapsed could have been saved.

I like to think that these ideas could be extended to the psychological, social, and economic arenas, in line with the notion that their mathematical and conceptual underpinnings have a great deal in common (Chapter 10). Rapid fluctuations in the state of a relationship with tiny changes in external circumstances, for example, may be a prelude to collapse, as happened with the on-again-off-again relationship of celebrity couple Marilyn Manson and Evan Rachel Wood, among many others. Such fluctuations could also signal a critical transition to the establishment of a new relationship and have been used as a literary device by romantic authors going back at least as far as the relationship between Elizabeth Bennett and Mr. Darcy in Jane Austen's *Pride and Prejudice.*

2. Greater Variance

Another significant statistical indicator of upcoming disaster is the occurrence of more extreme states. These can be easy to spot in relationships characterized by times of loving closeness interspersed with periods of violent argument. They are more difficult to observe in nature, although there is evidence for a greater frequency of extreme states in shallow lakes that are subjected to gradually rising levels of nutrients leached from adjacent farmlands. Such lakes are likely to suddenly "flip" from a clear to a turbid state—a process that would surely have shocked Stephen Forbes out of his belief in a long-term balance of nature. One warning that such a critical transition is about to occur is that phosphorus levels in the water begin to fluctuate more and more widely.

On a larger scale, we had better get used to the more extreme weather states brought on by global warming, such as that which drove the ash cloud from the Icelandic Eyjafjallajökull volcano across European airspace in April 2010. The ash cloud was blown in that direction because the prevailing westerly winds that would normally have carried it away from the continental landmass had been blocked by an unusual high-pressure weather system. According to simulations performed by Christophe Cassou of the European Centre for Research and Advanced Training in Scientific Computation, global warming will increase the frequency of such extreme blocking patterns.

3. Critical Slowing Down

Critical slowing down—which might also be described as "today is the same as yesterday"—is another symptom of a nearby catastrophic shift. A progressively slower ability to recover from small perturbations as a crisis approaches is a characteristic of reduced resilience.

The British humorist C. Northcote Parkinson, author of *Parkinson's Law or the Pursuit of Progress*, marvelously identified critical slowing down in ossifying business organizations as "injelititis." In the late stages of this "disease," according to Parkinson, "higher officials are plodding and dull, those less senior are active only in intrigue against each other, and the junior men are frustrated or frivolous. Little is being attempted. Nothing is being achieved."

In the final stages of the disease, a mean-spirited, self-centered, ignorant, and insecure top management team has gained control and things get even slower. The attitude in this terminal phase, Parkinson tells us, is this: "We rather distrust brilliance here. These clever people can be a dreadful nuisance, upsetting established routine and proposing all sorts of schemes that we have never seen tried." At this stage, smugness and apathy are the only remaining cultural anchors, and collapse is just around the corner.

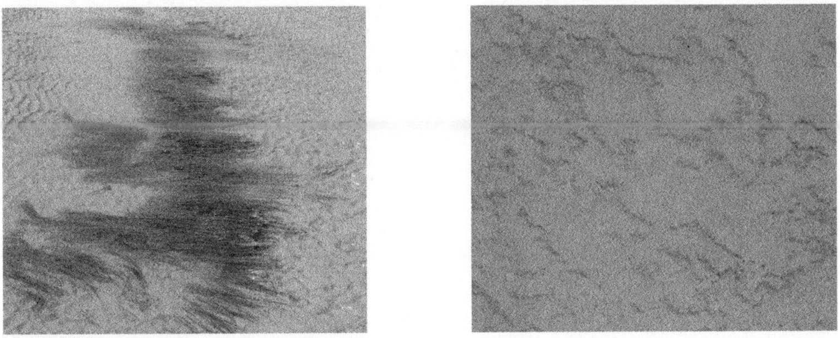

Figure 11.4. "Banding" in seagrass at Saint Efflam Bay, France. Photographs by T. van der Heide.

Critical slowing down is not always so funny. It has been mathematically demonstrated to be an early-warning sign for tipping points, and a careful examination of the history of the earth's climate shows that it has preceded eight of the most significant climate shifts, including the ends of various ice ages and the emergence of the present Holocene era a mere ten thousand years ago. It is now one of the principal symptoms that scientists are on the lookout for to see whether the global climate is shifting catastrophically from that era.

4. The Emergence of Particular Spatial Patterns

One way in which complex systems achieve stability is through the spontaneous emergence of spatial patterns, which develop when living conditions are improved by the presence of similar neighbors. One example from nature is the self-organized banding in beds of seagrass (Figure 11.4), which arises from the fact that seagrass plants can improve the growing conditions of nearby plants by attenuating currents and waves and reducing sediment resuspension. This process produces an environment with two alternative stable states: one where the plants are growing densely in lines whose direction is determined by the ambient currents, and one where they are not growing at all.

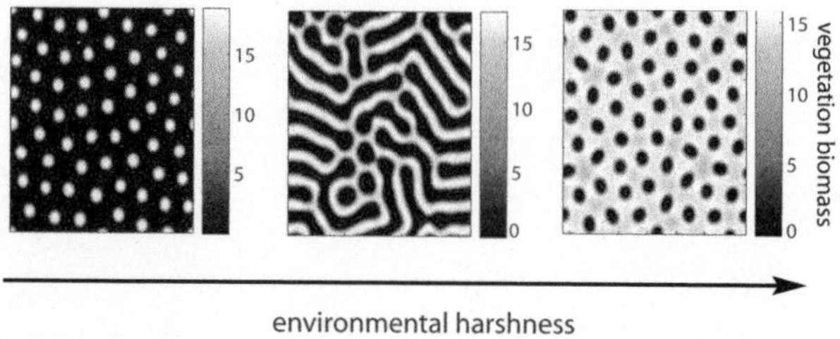

Figure 11.5. From left to right: progressively increasing patchiness of vegetation (dark areas) prior to desertification. Adapted from C. Valentin, J. M. d'Herbés, and J. Poesen, "Soil and Water Components of Banded Vegetation Patterns," *Catena* 37 (1999): 1–24.

Spatial patterns can emerge spontaneously in many different situations, from the patchy distribution of oases in deserts (where the community of plants forms a moisture-trapping microclimate) to the way in which different areas in cities become populated by different cultural groups. When the situation changes, changes in the spatial patterns can be a signal of upcoming critical transitions.

It happened in the English village where I lived, which had been stably populated by generations of the same families. There was even a "lord of the manor" to oversee proceedings, not to mention an ancient castle. Then came an upswing in the economy and a fashion for living in old country cottages, which raised the prices of those cottages until they were beyond the means of the younger generation. A flood of newcomers rapidly changed the pattern of the village, with the desirable old cottages now inhabited by newcomers (including myself) and the locals pushed to the periphery.

In nature, changes in pattern can signal large-scale disaster. If the climate in a semi-arid region becomes dryer, for example, the pattern of vegetation slowly becomes patchier in a very predictable way (Figure 11.5), until it reaches a critical point where the remaining vegetation suddenly dies and the region is left barren.

5. Increasing Skewness in the Distribution of States

Changes in pattern, critical slowing down, increasing scatter, and flickering all provide hints of upcoming catastrophe, but there is yet one more—increasing skewness.

"Skewness" means *asymmetry*, as in the increasing divide between the rich and the poor in many countries that produces a distorted, asymmetric distribution of wealth. Similarly distorted, asymmetric distributions are seen in nutrient concentrations for up to five years before clear shallow lakes "flip" to a turbid state, and they were seen in the distribution of vegetation in North Africa 5,500 years ago prior to the formation of the Sahara Desert.

Skewness is a warning sign, but like so many other weak signals, it can be an ambiguous warning sign. Close examination of the Sahara data, for example, reveals that the change in skewness might have arisen from statistical error.

This emphatically does *not* mean that we should ignore such warning signs. The desertification of the Sahara region was also preceded by other warning signs, such as an increased patchiness of vegetation. Taken individually, different warning signs are ambiguous. Taken together, they have an overwhelmingly cumulative force.

Science can't make absolute predictions, no matter how much politicians and journalists demand it. As the Nobel laureate Niels Bohr once pointed out, "Prediction is very difficult, especially about the future." All that science can do is to provide the evidence and calculate the odds, using multiple warning signs where possible to produce the best estimate of those odds. The rest is up to us.

With respect to global warming, all of the warning signs point to evidence that it is happening, and there is at least a 90 percent chance that humankind's activities are making a substantial contribution. In the face of these odds, the least we should do is to take appropriate measures to prepare ourselves for the possible eventuality of a catastrophically different world—as the toads did before the L'Aquila earthquake, and as business advisers recommend—in order to produce the best chance for survival. We can do no more, but we should do no less.

If the toads could get it right, surely we should be able to as well, even without their special sense of danger.

Summary: The Future of Forecasting

The message of this book is simple: we *can* learn to tell when disasters or other sudden changes are imminent, both in our personal lives and in our economies, our societies, our ecosystems, and the totality of our global environment.

The warning signs can come from classical science, or they may have been uncovered by the latest computer-based analyses. They can be clear-cut, or they may be ambiguous and difficult to read. But they are there. As the stories in this book have revealed, we already know enough about them to help us anticipate change, to prepare ourselves for it, and to ameliorate its effects even if we cannot avoid them entirely.

These are main warning signs for upcoming sudden and sometimes disastrous change in personal, social, economic, physical, or environmental circumstances:

Unacceptable buildup of stress
Concentration of stress at weak points
Potential for uncontrolled runaway effects
Loss of resilience (the ability to recover rapidly from small disturbances)
Increasing swings between different states

Increasing occurrence of extreme states
Changes in pattern

Just thinking about these signs in terms of our own personal relationships is sufficient to show their importance in warning us of disaster ahead. In the hands of scientists, they are becoming an important quantitative tool to help forecast disasters of all kinds. We can only hope that, like the toads of L'Aquila, we learn to read and heed them in time.

This book is part of an ongoing project—to understand how science can help us plan for a better future, not just in the practical things of life, but through contributing to our understanding of how societies, economies, and ecosystems function. For a continuing discussion, with continuously updated examples, see my website at www.lenfisherscience.com.

NOTES

INTRODUCTION: HOW DO TOADS PREDICT EARTHQUAKES?

xi **"The clever men at Oxford":** Kenneth Grahame, *The Wind in the Willows*, ch. 10, available as a Project Gutenberg e-book at: http://www.gutenberg.org/files/289/289.txt.

xi **the toads . . . left their traditional breeding grounds:** R. A. Grant and T. Halliday, "Predicting the Unpredictable: Evidence of Pre-seismic Anticipatory Behavior in the Common Toad," *Journal of Zoology* 281 (2010): 263–271.

xi **a violent earthquake hit the region:** BBC News, "Powerful Italian Quake Kills Many," April 6, 2009, http://news.bbc.co.uk/1/hi/7984867.stm; National Environment Research Council, "L'Aquila Earthquake: A Year On," PlanetEarthOnline, June 28, 2010, http://planetearth.nerc.ac.uk/features/story.aspx?id=753.

xi **killing more than three hundred people:** One upshot of this human tragedy, with serious implications for the use of science in prediction, was the decision of the L'Aquila public prosecutor to indict six of Italy's top seismologists on a charge of manslaughter for failing to predict the earthquake; see Lisa Zyga, "Italian Scientists Who Failed to Predict L'Aquila Earthquake May Face Manslaughter Charges," PhysOrg.com, June 24, 2010, http://www.physorg.com/news196622867.html.

xi **Rachel Grant happened to be studying the . . . toads:** Grant and Halliday, "Predicting the Unpredictable."

xi **She was understandably annoyed:** "Toads 'Predict Earthquakes,'" post by Janet Fang on The Great Beyond (*Nature* magazine blog), March 31, 2010, http://blogs.nature.com/news/thegreatbeyond/2010/03/toads_predict_earthquakes.html.

xi **many anecdotal reports over the centuries:** Helmut Tributsch, *When the Snakes Awake: Animals and Earthquake Prediction* (Cambridge, Mass.: MIT Press, 1984); Rupert Sheldrake, *Dogs That Know When Their Owners Are Coming Home* (New York: Three Rivers Press, 2000).

xi **use the behavior of toads or other animals as early-warning signs for earthquakes:** There have been occasional reports of people noting unusual animal behavior and acting on it to save lives. One story with wide circulation is that, in the winter of 1975, Chinese officials ordered the evacuation of the city of Haicheng in northeastern Liaoning province the day before a 7.3 magnitude earthquake, based on reports of unusual animal behavior and changes in groundwater levels.

Unfortunately, this story is an urban myth that seems to have originated from those same officials. The earthquake was certainly predicted and is believed by reputable scholars to have been "the only major earthquake ever to have been predicted," according to Kelin Wang, Qi-Fu Chen, Shihong Sun, and Andong Wang ("Predicting the 1975 Haicheng Earthquake," *Bulletin of the Seismological Society of America* 96 [2006]:

757–795). These authors, however, after carefully examining the records, concluded that the predictions were based on conventional geophysical methods and that observations of unusual animal behavior were an insignificant factor at best.

xi **at least we may be able to discover any physical signals that the animals are responding to:** Joseph L. Kirschvink ("Earthquake Prediction by Animals: Evolutionary and Sensory Perception," *Bulletin of the Seismological Society of America* 90 [2000]: 312–323) gives a detailed account of the most likely signals. He also points out that, if we can relate the animal behavior to any measurable variable, then we will generally be better off simply measuring the variable and forgetting about the animal behavior, which is subject to its own range of variability.

xi ***critical transitions*:** Marten Scheffer, *Critical Transitions in Nature and Society* (Princeton, N.J.: Princeton University Press, 2009).

xiv **intolerable stresses may provoke a sudden collapse:** The Pentagon labels such disasters as BOOBS—Bolts Out of the Blue. The acronym was invented to describe the possibility of surprise attacks by nuclear missiles during the cold war and is now used in Pentagon internal memoranda about potential terrorist threats (Robert L. Butterworth, "Out of Balance: Will Conventional ICBMs Destroy Deterrence?" *Aerospace Power Journal* [Fall 2001], http://www.airpower.maxwell.af.mil/airchronicles/apj/apj01/fal01/butterworth.html).

xv **a remarkable set of universal early-warning signs:** Marten Scheffer et al., "Early-Warning Signals for Critical Transitions," *Nature* 461 (2009): 53–59.

CHAPTER 1: DO ANIMALS HAVE CRYSTAL BALLS?

3 **"the entire universe has been neatly divided":** The third novel in Terry Pratchett's phenomenally successful "Discworld" series, *Equal Rites* (London: Corgi Books, 1987) is the first to feature Granny Weatherwax, whose knowledge of "headology" and understanding of the power of belief mark her as an appropriate presiding muse for the theme of this chapter.

3 **the Greek city of Helike:** For a compendium of classical descriptions of this event, see Dora Katsonopoulou and Steven Soter, "Appendix A: Principal Ancient Sources on Helike," The Lost Cities of Ancient Helike, http://www.helike.org/sources2.shtml.

3 **Five hundred years later, the ruins could still be seen:** The Greek traveler Pausanias wrote in his *Description of Greece* (http://www.helike.org/sources2.shtml), published around AD 160, that "the ruins of Helike are also visible, but not so plainly now as they were once, because they are corroded by the salt water."

3 **Roman tourists would sail above them:** The Roman poet Ovid (43 BC–AD 17 or 18) writes in *Metamorphoses* that, "should you seek Helike and Buris cities of Achaia, you will find them beneath the waves, and the sailors are still wont to point out [these] leveled towns, with their walls buried under water" (available as a Project Gutenberg e-book at: http://www.gutenberg.org/files/26073/26073-8.txt).

3 **the story of Atlantis may have been based on these ruins:** The story appears in Plato's *Timaeus* (available as a Project Gutenberg e-book at: http://www.gutenberg.org/files/1572/1572.txt), written thirteen years after Helike sank beneath the waves—an event that Plato would have known about. Plato writes about Atlantis as though it were

a real place, but his student Aristotle believed that Plato had made it up (possibly on the model of Helike, although there is no real evidence either way) simply to illustrate a point. "The man who dreamed it up," said Aristotle, "made it vanish" (quoted in Strabo's *Geography*, available at UnMuseum: The Lost Continent of Atlantis, http://www.unmuseum.org/atlantis.htm; Aristotle's original writings on the subject have long since disappeared). People still keep trying, though; see, for example, Peter Popham, "Architect's Mission in Cyprus: One Man's Quest to Find Atlantis," *The Independent*, August 24, 2005, http://www.independent.co.uk/news/world/europe/architects-mission-in-cyprus-one-mans-quest-to-find-atlantis-504040.html.

3 **Helike disappeared:** See, for example, Maurice L. Schwartz and Christos Tziavos, "Geology in the Search for Ancient Helice," *Journal of Field Archaeology* 6 (1979): 243–252. Even in 1979 there was still no real clue as to the whereabouts of the ancient city. Note also the spelling. "Helike" is pronounced *he–lick–kee*, but often spelled "Helice." Confusing.

3 **The city was not rediscovered until 2001:** Thanks mainly to the efforts of the Greek archaeologist Dora Katsonopoulou, who set up the Helike project in 1988 (http://www.helike.org/paper.shtml).

3 **the first description of animals "predicting" an earthquake:** This description is often mistakenly attributed to the Greek historian Diodorus Siculus. Helmut Tributsch (*When the Snakes Awake*) seems to have started the trend, and subsequent authors have followed like sheep.

4 **animals fleeing for safety before the disastrous Asian tsunami:** Maryann Mott, "Did Animals Sense Tsunami Was Coming?" *National Geographic News*, January 4, 2005, http://news.nationalgeographic.com/news/2005/01/0104_050104_tsunami_animals.html.

4 **Some people believe that animals have a "sixth sense":** See, for example, Bill Schul, *The Psychic Power of Animals* (1977; reprint, e-reads.com, 2003).

4 **claimed ability of some dogs to "know" when their owners are coming home:** Rupert Sheldrake, *Dogs That Know When Their Owners Are Coming Home* (New York: Three Rivers Press, 2000).

4 **When the psychologist Richard Wiseman and his colleagues investigated the claim:** Richard Wiseman, Matthew Smith, and Julie Milton, "Can Animals Detect When Their Owners Are Returning Home? An Experimental Test of the 'Psychic Pet' Phenomenon," *British Journal of Psychology* 89 (1998): 453–462, https://uhra.herts.ac.uk/dspace/bitstream/2299/2285/1/902380.pdf.

4 **the person filming the dog's behavior . . . may have given inadvertent signals:** Wiseman requested a copy of the original film from the producers to check this point out, but was told that it had been "lost."

5 **Housed in a German aquarium [Paul the octopus]:** That facility is the Aquarium Sea Life Centre in Oberhausen (http://www.sealifeeurope.com/local/index.php?loc=oberhausen).

5 **[Paul the octopus] correctly predicted the result of every match:** CNN, "Paul the Octopus Retires with World Cup Record Intact," July 13, 2010, http://edition.cnn.com/2010/WORLD/europe/07/13/germany.paul.the.octopus/#fbid=53Wf4CG

wRmw&wom=false. See also the Wikipedia article (http://en.wikipedia.org/wiki/Paul_the_octopus) for many interesting and well-referenced additional bits of information.

5 **some fans calling for the octopus to be cooked and eaten:** Erik Kirschbaum, "German Fans Want Revenge Grilling of Oracle Octopus," Reuters, July 8, 2010, http://www.reuters.com/article/idUSTRE6673P220100708.

5 **Iranian president Mahmoud Ahmadinejad . . . accused the West of using the octopus:** "Mahmoud Ahmadinejad Attacks Octopus Paul," *Daily Telegraph* (United Kingdom), August 4, 2010.

6 **toads . . . may have been sensitive to the increase in very low-frequency radio emissions:** Grant and Halliday, "Predicting the Unpredictable."

6 **The same seismologists have pointed out:** Kirschvink, "Earthquake Prediction by Animals: Evolutionary and Sensory Perception."

6 **Croesus . . . decided to investigate [the Oracles]:** Herodotus, *The Histories*, available as a Project Gutenberg e-book at: http://www.gutenberg.org/cache/epub/2707/pg2707.txt.

7 **[Croesus was] boiling up a tortoise and a lamb:** The classicist Marcia Dobson has challenged the veracity of Herodotus's account ("Herodotus 1.47.1 and the Hymn to Hermes: A Solution to the Test Oracle," *American Journal of Philology* 100 [1979]: 349–359).

7 **the famous Oracle . . . in the Greek town of Delphi:** William J. Broad, *The Oracle: Ancient Delphi and the Science Behind Its Lost Secrets* (New York: Penguin, 2007).

7 **She would then sniff the volcanic vapors:** The story of sniffing gases from a crack in the ground was originally reported by the Greek historian Plutarch in *Moralia*, vol. 5 (Cambridge, Mass.: Harvard University Press, 1936), 498–499, and accepted on trust until 1950, when it was ridiculed by the French archaeologist Pierre Amandry in *La Mantique apolliniene Delphes* (Paris: Boccard, 1950). Amandry correctly pointed out that there is now no chasm under the Temple of Apollo at Delphi, and in any case the region is not volcanic.

In a further twist to the tale, the geologist Jelle de Boer teamed up with the archaeology teacher John Hale in the mid-1990s to reexamine the evidence (Jelle Z. de Boer and John R. Hale, "The Geological Origins of the Oracle at Delphi, Greece," *Geological Society, London, Special Publications* 171 [2000]: 399–412). They concluded that there is ample evidence to support Plutarch's original description. According to these authors, "the tectonic data show that the Delphi oracle site is located at the intersection of a major WNW-ESE, southward dipping fault zone and a swarm of minor NNW-SSE trending, predominantly east dipping fractures," which could certainly have opened up gas-releasing cracks in the pre-Christian era.

Among these gases would have been ethylene, which was still in use as an anesthetic gas in the 1930s (Isabella C. Herb and Hubbard Woods, "The Present Status of Ethylene," *British Journal of Anaesthesia* 11 [1934]: 66–71). One of the early stages of anesthesia is "excited and delirious activity," which could account for the Oracle's performance—and also for the incomprehensibility of her ravings. Another gas that would have been present was hydrogen sulfide ("rotten egg gas"), which is more lethal than hydrogen cyanide. It would be interesting to know what the average life span of an Oracle was.

Individuals who claim to be able to visualize future events while they are in an ecstatic, trancelike state have been a feature of many cultures. In some Amerindian tribes, the state was induced by peyote; the classic account is Weston La Barre, *The Peyote Cult*, 5th enlarged ed. (Norman: University of Oklahoma Press, 1989), first published in 1938. For an up-to-date account, see Edward F. Anderson, *Peyote: The Divine Cactus*, 2nd ed. (Tucson: University of Arizona Press, 1996).

Music has been used to induce a similar state. Edward Fitzgerald describes an example in *The Golden Bough: Adonis, Attis, Osiris*, vol. 1 (London: Macmillan, 1936), 52. According to Fitzgerald, "At Jerusalem the regular clergy prophesied to the music of harps, of psalteries, and of cymbals; and it appears that the irregular clergy also, as we may call the prophets, depended on some such stimulus for inducing the ecstatic state."

8 **James Randi . . . has a standing offer of $1 million:** Jeff Wagg, "One Million Dollar Paranormal Challenge," October 24, 2008, James Randi Educational Foundation, http://www.randi.org/site/index.php/1m-challenge.html.

8 **the Australian sheep-goat scale:** This scale was devised by the Adelaide psychologist Michael Thalbourne (M. A. Thalbourne and P. S. Delin, "A New Instrument for Measuring the Sheep-Goat Variable: Its Psychometric Properties and Factor Structure," *Journal of the Society for Psychical Research* 59 [1993]: 172–186). It is named after the New Testament story that recounts how the Son of Man "will separate the people one from another as a shepherd separates the sheep from the goats" (Matthew 25:31–33).

8 **the Oracle's supposed ability to see at least dimly into the future:** This ability was also shared by elevators in Douglas Adams's *The Restaurant at the End of the Universe* (London: Pan Books, 1980), the second book in his "Hitchhiker's Guide to the Galaxy" series: "Modern elevators . . . operate on the curious principle of 'defocused temporal perception.' In other words, they have the capacity to see dimly into the immediate future, which enables the elevator to be on the right floor to pick you up even before you knew you wanted it" (p. 41).

9 **Changizi's research:** Mark A. Changizi, Andrew Hsieh, Romi Nijhawan, Ryota Kanai, and Shinsuke Shimojo, "Perceiving the Present and a Systematization of Illusions," *Cognitive Science* 32 (2008): 459–503. This is a fairly technical article; for a popular and accessible presentation, see "Crystal (Eye) Ball: Visual System Equipped with 'Future Seeing Powers,'" *Science Daily*, May 16, 2008, http://www.sciencedaily.com/releases/2008/05/080515145356.htm.

9 **"extraordinary claims require extraordinary proof":** Carl Sagan's dictum appeared in *Cosmos* (New York: Ballantine Books, 1985). It is based on an argument advanced by the Scottish philosopher David Hume two hundred years earlier ("Of Miracles," in *An Enquiry Concerning Human Understanding*, section X, part 1, available as a Project Gutenberg e-book at: http://www.gutenberg.org/dirs/etext06/8echu10.txt).

9 **the Cambridge philosopher C. D. Broad:** C. D. Broad, "The Experimental Establishment of Telepathic Precognition," *Philosophy* 19 (1944): 261–275.

10 **some of the most significant results were almost certainly fabricated:** For the full story, see Andrew M. Colman, *Facts, Fallacies, and Frauds in Psychology* (New York: Unwin Hyman, 1988), 175–180.

10 **four of which were identified by Broad:** C. D. Broad, "The Philosophical Implications of Foreknowledge," *Proceedings of the Aristotelian Society* (supplement) 16 (1937): 177–209.

10 ***Cryptomnesia:*** Cryptomnesia has frequently been implicated when people claim to have had "past lives." One twenty-three-year-old woman named Jan claimed under live hypnosis on British television to have had a previous life in the sixteenth century. She even described a famous 1566 witchcraft trial and acquittal of one Joan Waterhouse, a famous witch of Chelmsford. Unfortunately, Jan gave the wrong date of 1556—a mistake that had slipped into a Victorian reprint of the trial proceedings, of which there were only two copies. One of these was displayed in the British Museum, where Jan had almost certainly seen it, but then forgotten that she had seen it. See the definition of cryptomnesia at the Mystica, http://www.themystica.com/mystica/articles/c/cryptomnesia.html.

11 **the experience of those who have had "premonitions":** See, for example, the anecdotal accounts in Keith Hearne and Jane Henry, "Precognition and Premonitions," in *Parapsychology: Research on Exceptional Experiences*, edited by Jane Henry (London: Routledge, 2004), 108–113.

11 **"Say the odds are a million to one":** Robert Todd Carroll, "Law of Truly Large Numbers (Coincidence)," *The Skeptic's Dictionary*, http://skepdic.com/lawofnumbers.html.

11 **Aristotle dismissed precognition:** Aristotle, *On Prophesying by Dreams*, Internet Classic Archives, http://classics.mit.edu/Aristotle/prophesying.html. One of Aristotle's arguments is that "it is absurd to combine the idea that the sender of such dreams should be God with the fact that those to whom he sends them are not the best and wisest, but merely commonplace persons."

11 **Freud himself ridiculed the idea that dreams could reveal the future:** Sigmund Freud, *Dream Psychology: Psychoanalysis for Beginners*, available as a Project Gutenberg e-book at: http://www.gutenberg.org/files/15489/15489-h/15489-h.htm.

11 **There is, in fact, *no* scientific evidence that we can directly visualize the future:** In other words, no replicable demonstration has been achieved. See Christopher Scott, "Paranormal Phenomena: The Burden of Proof," in *The Oxford Companion to the Mind*, edited by Richard L. Gregory (Oxford: Oxford University Press, 1987), 578–581.

12 **his remarkable treatise:** *De Divinatione*, translated by William Falconer (Cambridge, Mass.: Harvard University Press, 1964), 399–465.

12 **Cicero's argument:** This is my simplified paraphrase of Cicero's rather convoluted original.

12 **divining the future is logically impossible:** My own minor contribution to this philosophical problem has been to demolish an argument that some conditional statements (statements of the form "if . . . then . . .") can be reliably declared to be true or false before the event (Len Fisher, "Truth Conditions for Counterfactuals: Limits to Lewis's Limit Assumption," master's thesis, University of Bristol, 1992).

12 **the prestigious Niels Bohr Institute in Copenhagen:** If my experience of the institute is anything to go by, the idea that emerged from it of a tiny particle being able to hide itself has some merit: The last time I was there the whole institute managed to hide itself from my sight as I cycled around fruitlessly searching for it at the address given.

12 The theory was brought to public attention by the *New York Times*: Dennis Overbye, "The Collider, the Particle, and a Theory About Fate," *New York Times*, October 12, 2009, http://www.nytimes.com/2009/10/13/science/space/13lhc.html?_r=2&em=&pagewanted=all. For a somewhat less sensational account, see Richard Webb, "Time-Traveling Higgs Sabotages the LHC. No, Really," *New Scientist*, October 13, 2009, http://www.newscientist.com/blogs/shortsharpscience/2009/10/is-a-time-travelling-higgs-sab.html. For the original references and ongoing technical discussion (warning: high-level mathematics and physics required), see Iain Stewart, "An Iterated Search for Influence from the Future on the Large Hadron Collider," arXive:0712.0715v2 (hep-ph), December 15, 2007, and Holger Nielsen and Masao Ninomiya, "Card Game Restriction in LHC Can Only Be Successful!," arXiv:0910.0359v3 (physics.gen-ph), October 23, 2009.

12 The "particle" was the Higgs boson: The existence of the Higgs boson was postulated by Professor Peter Higgs of Edinburgh University in 1964 in a paper that was rejected by the editors of *Physics Letters* (based at CERN!) as being "of no obvious relevance to physics" (!) ("Peter Higgs: The Man Behind the Boson," *Physics World*, July 10, 2004, http://physicsworld.com/cws/article/print/19750). The paper was eventually accepted by *Physical Review Letters*, after the addition of a paragraph, and published later in the same year (Peter W. Higgs, "Broken Symmetries and the Masses of Gauge Bosons," *Physical Review Letters* 13 [1964]: 508–509).

The Higgs boson is sometimes known as "the God particle"—a title used by the Nobel laureate Leon Lederman when he wrote a popular book about it, *The God Particle: If the Universe Is the Answer, What Is the Question?* (Boston: Houghton Mifflin, 1993). There have been many fanciful guesses about the origin of the term, but when I asked Professor Lederman himself, his answer was: "The only rationale I can recall is that we were dealing with a very important particle." When I inquired whether he had coined the term, or whether it had been in the air at the time, he replied, "When you are desperate, all ideas are 'in the air.'" I was left with the feeling that he was rather embarrassed by the term.

13 the Texas-based Superconducting Super Collider: The history of this ill-fated project is given on the high-energy physics hep.net website: http://www.hep.net/ssc/new/history/appendixa.html.

14 Time will tell: For an accessible summary of the current state of the physics and philosophy of time, see Nature of Time, http://www.timephysics.com/nature-of-time.html.

14 claims that . . . quantum mechanics may make . . . precognition possible: See, for example, Paul Karl Hoiland, "Scientific Grounds for Precognition," 2003, cogprints.org/2851/1/SCIENTIFIC_GROUNDS_FOR_PRECOGNITION.pdf.

14 such effects are unprovable: R. L. Schafir, "Unprovability of a 'Precognition' Effect in Quantum Mechanics," *Foundations of Physics Letters* 9 (1996): 91–101.

14 the existence of parallel universes: Mark Tegmark, "Parallel Universes," in *Science and Ultimate Reality: From Quantum to Cosmos*, edited by J. D. Barrow, P. C. W. Davies, and C. L. Harper (Cambridge: Cambridge University Press, 2003), space.mit.edu/home/tegmark/multiverse.pdf. This is a very readable and informative article for the layperson.

14 **Or perhaps science fiction:** The logical absurdities inherent in precognition have been explored by many science fiction writers, starting with H. G. Wells in *The Time Machine,* written in 1895. Wells rather avoided talking about paradoxes, but his ideas were used by later writers as a basis for exploring this theme. Philip K. Dick, for example, wrote a series of stories about mutant humans, called "precogs," who were able to see into the future. In *The Golden Man,* one precog becomes capable of seeing ahead right up to the time of his own death, and thus he becomes completely nonhuman, an animal compelled to follow a preordained path with no choice whatsoever. Other notable examples are John Wyndham's marvelous short story "Chronoclasm," Isaac Asimov's *The End of Eternity,* C. S. Lewis's *Chronicles of Narnia,* and J. K. Rowling's *Harry Potter and the Prisoner of Azkaban,* where the grandfather paradox is resolved in ingenious fashion. One might also include Douglas Adams's *The Restaurant at the End of the Universe,* in which an enormous five-star restaurant rocks endlessly back and forth across the end of time, requiring the invention of a new language to describe the confused relationship between past, present, and future.

CHAPTER 2: THE FUTURE ECLIPSED

15 **"The clouds I can handle":** Stephenie Meyer's *Eclipse* (London: Atom, 2007) is the third novel in the "Twilight" series.

15 **two Chinese court astronomers called Hsi and Ho:** The fate of the astronomers is recorded in the ancient Chinese document *Shu Ching,* which reported that "the Sun and Moon did not meet harmoniously." There were several eclipses within the historical period that this document covers; this is the most likely one (*Shu Ching: Book of History,* edited by F. Max Muller and translated by James Legge [New Delhi: Motilal Banarsidass, 1988]).

15 **they even had a primitive planetarium:** F. Crawford Brown, "The Eclipse in China," *Popular Astronomy* 39 (December 1931): 567–572.

15 **According to the ancient Chinese document *Shu Ching*:** The version cited here was translated by the American astronomer and historian Robert Newton from a French version, which had in turn been translated from the Chinese (Robert R. Newton, *Ancient Astronomical Observations and the Accelerations of the Earth and Moon* [Baltimore: Johns Hopkins University Press, 1970], 62–65). Newton also raises the point that Hsi and Ho may not have been real people at all and that the names may derive from that of the minor deity Hsi-Ho.

16 **a full solar eclipse [only occurs] once every four hundred years:** Matthew Cavagnaro, "Moon Dance," March 15, 2004, NASA, http://www.nasa.gov/missions/solarsystem/f-eclipse.html.

16 **a dragon trying to eat the Sun:** Ibid.

16 **This fascinating story [was] . . . probably . . . fabricated around AD 300:** Newton, *Ancient Astronomical Observations and the Accelerations of the Earth and Moon.* See also R. R. Newton, "Two Uses of Ancient Astronomy," *Philosophical Transactions of the Royal Society of London* A276 (1974): 99–110, http://www.pereplet.ru/gorm/atext/newton2.htm.

16 **ancient civilizations took eclipses very seriously as harbingers of important events:** So seriously, in fact, that an eclipse once stopped a war that had been going on for five years.

The war was between the Medes, who lived in the northwestern portions of present-day Iran, and the Lydians, who occupied what is now western Turkey. The final battle took place on the banks of the river Halys (now in northern Turkey) on May 28, 585 BC. (Trivia buffs may care to note that Croesus, the future fabulously wealthy king of the Lydians—from whom we get the expression "as rich as Croesus"—was just five years old at the time.) The Greek historian Herodotus (*The Histories*, vol. 1, translated by G. C. Macaulay, available as a Project Gutenberg e-book at: http://www.gutenberg.org/cache/epub/2707/pg2707.txt) described what happened when the eclipse put in its appearance:

> In the sixth year a battle took place in which it happened, when the fight had begun, that suddenly the day became night. And this change of the day Thales the Milesian had foretold. . . . The Lydians however and the Medes, when they saw that it had become night instead of day, ceased from their fighting and were much more eager both of them that peace should be made between them.

Thales has been called "the father of science" (Charles Singer, *A Short History of Science to the Nineteenth Century* [Oxford: Clarendon Press, 1941], 35). His prediction shows just how much the science of astronomy had progressed since the days of Hsi and Ho, mainly driven by the wish to be able to predict eclipses and other unusual heavenly events, which were still seen as highly significant. Needless to say, the warring parties had no knowledge of his prediction.

Superstitious fear of eclipses did not always produce such a fortunate outcome. When Athenian troops were besieging the Sicilian city of Syracuse during the long-running Peloponnesian War with Sparta, the siege had not gone well, and the Athenian commanders had just decided on a tactical withdrawal, when a lunar eclipse occurred on August 27, 413 BC. What happened next is described by Thucydides in Chapter 23 of his *History of the Peloponnesian War* (available as a Project Gutenberg e-book at: http://www.gutenberg.org/files/7142/7142.txt): "Most of the Athenians, deeply impressed by this occurrence, now urged the generals to wait; and Nicias [their commander], who was somewhat over-addicted to divination and practices of that kind, refused from that moment even to take the question of departure into consideration, until they had waited the thrice nine days prescribed by the soothsayers."

It was the turning point in the war. While the Athenians waited, the Syracusans managed to blockade the port where the Athenian ships were moored, so that they could no longer escape. Within weeks, the entire Athenian force (an appreciable proportion of the city's total manpower) was wiped out.

16 **the astrophysicist David Dearborn:** A researcher at the Lawrence Livermore National Laboratory in Berkeley, California, Dearborn is a specialist in "archaeoastronomy." The quote is from Noel Wanner, "The Sun-Eating Dragon: Eclipse Stories, Myths, and

Legends," Solar Eclipse, http://www.exploratorium.edu/eclipse/dragon.html. See also K. Kris Hirst, "Archaeoastronomy: An Interview with David Dearborn," http://archaeology.about.com/cs/archaeoastronomy/a/dearborn.htm.

17 **the Higgs boson is thought to have first put in an appearance [when the universe was being born]:** "Missing Higgs," European Organization for Nuclear Research (CERN), http://public.web.cern.ch/public/en/science/higgs-en.html.

17 **a shower of other particles:** Internet Encyclopedia of Science, "Higgs Boson," http://www.daviddarling.info/encyclopedia/H/Higgs_boson.html.

18 **The ancient civilizations . . . used correlations:** Erica Reiner, "Babylonian Celestial Divination," in *Ancient Astronomy and Celestial Divination*, edited by Noel M. Swerdlow (Cambridge, Mass.: MIT Press, 2000), 22–23; Francesca Rochberg, "Heaven and Earth: Divine-Human Relations in Mesopotamian Celestial Divination," in *Magic in History: Prayer, Magic, and the Stars in the Ancient and Late Antique World*, edited by Scott Noegel, Joel Walker, and Brannon Wheeler (University Park: Pennsylvania State University Press, 2003), 180.

18 **the "omen series":** The "omen series" was partly published by Erica Reiner and David Pingree as a set of fascicles entitled *Babylonian Planetary Omens.* The first twenty-two books contain predictions for king and country derived from lunar phenomena and the moon's general appearance: horns, halos, conjunctions, and especially eclipses. Books 23 to 37 cover solar phenomena such as halos, colors, and eclipses. Books 37 to 49 treat meteorological phenomena like thunder, lightning, rainbows, and winds, as well as earthquakes. The final books, 50 to 70, discuss the positions of the planets and the fixed stars.

In addition to eclipses, the relative positions of the planets and stars were also given weight. Each unusual arrangement (including some that were astronomically impossible and are still puzzling scholars) was given its own particular predictive significance. The Cal Tech historian Noel Swerdlow tells me that

> the omens are not [necessarily] based upon an actual record of eclipses and other phenomena, just on the sort of phenomena that can (and in some cases cannot) occur. They were probably not put together by making empirical correlations between phenomena in the heavens and actual events on earth. [*Note:* As a scientist, I find this hard to believe. Surely *some* of the correlations must have been based on observation, even if there is no written evidence of this.] Some of the relations between the protasis ("if" clause) and apodosis ("then" clause) of omens depend upon plays on similar words, some exhaust every possibility of some kind of phenomenon, say, an eclipse from different directions, and what will happen, say, a king from different regions will have misfortune. [So] the omen series is neither an empirical nor a historical record. It is something else, and I don't think just what it is, or all that it is, is yet understood (and may never be).

18 **heavenly events . . . [and the] earthly events with which they were thought to be correlated:** These pairs consisted of a *protasis* (if this happens, then . . .) and *apodosis* (a forecast event), which together are described by the technical term *omen.*

18 **the pseudo-science of astrology:** According to Paul Thagard ("Why Astrology Is a Pseudoscience," *Proceedings of the Biennial Meeting of the Philosophy of Science Association* [1978]: 223–224), "a theory or discipline which purports to be scientific is pseudoscientific if and only if:

1. It has been less progressive than alternative theories over a long period of time, and faces many unsolved problems; but
2. The community of practitioners makes little attempt to develop the theory towards solutions of the problems, shows no concern for attempts to evaluate the theory in relation to others, and is selective in considering confirmations and disconfirmations."

Astrology fits on both counts.

The Berkeley physicist Shawn Carlson decided to test astrology scientifically ("A Double-Blind Test of Astrology," *Nature* 318 [1985]: 419–425), and specifically to test astrologers' claim that they can predict "character" from birth sign. He invited leading astrologers to help him set up the experiments. His conclusion was that

> despite the fact that we worked with some of the best astrologers in the country, recommended by the advising astrologers for their expertise in astrology and their ability to use CPI [California Personality Inventory, a standardized list of personality traits], despite the fact that every reasonable suggestion made by the advising astrologers was worked into the experiment, despite the fact that the astrologers approved the design and predicted 50 percent as the "minimum" effect they would expect to see, astrology failed to perform at a level better than chance. Tested using double-blind methods, the astrologers' predictions proved to be wrong. Their predicted connection between the positions of planets and other astronomical objects at the time of birth and the personalities of test subjects did not exist.

Yet many people still believe in astrology! What more proof do they want?

The answer, of course, is that they don't want *any*. The post hoc fallacy has them in its grasp—sometimes reinforced by a misunderstanding or misuse of statistics. If you do enough tests, a few spurious correlations are bound to emerge, as demonstrated by a group of biologists in an amusing study on the inhabitants of Ontario. By juggling recognized statistical tests, the biologists managed to "prove" that aspirin increased the mortality of patients born under the astrological sign of Gemini or Libra between 1990 and 1999 (Peter C. Austin et al., "Testing Multiple Statistical Hypotheses Resulted in Spurious Associations: A Study of Astrological Signs and Health," *Journal of Clinical Epidemiology* 59 [2006]: 964–969).

18 **some commentators liken [prediction methods based on correlations] to astrology:** Spyros Makridakis and Nassim Taleb, "Decision Making and Planning Under Low Levels of Predictability," *International Journal of Forecasting* 25 (2009): 716–733. As Makridakis and Taleb point out, the track record of modern forecasters is almost as dismal as that of their ancient Mesopotamian counterparts. "Who," they ask,

predicted the subprime and credit crunch crises, the Internet bubble, the Asian contagion, the real estate and savings and loan crises, the Latin American lending calamity, and other major disasters? In business, who "predicted" the collapse of Lehman Brothers, Bear Stearns, AIG, Enron or WorldCom (in the USA), and Northern Rock, Royal Bank of Scotland, Parmalat or Royal Ahold (in Europe); or the practical collapse of the entire Iceland economy? In finance, who predicted the demise of LTCM and Amaranth, or the hundreds of mutual and hedge funds that close down every year after incurring huge losses? And these are just the tip of the iceberg.

18 **a spike in the gold price:** Frank Ahrens, "The Frightening Spike in the Price of Gold," *Washington Post*, September 24, 2009, http://voices.washingtonpost.com/economy-watch/2009/09/gold_is_atnear_a_historic.html.

19 **It's a belief that has served us well throughout evolution:** Kevin R. Foster and Hanna Kokko, "The Evolution of Superstitious and Superstition-Like Behavior," *Proceedings of the Royal Society B* 276 (2009): 31–37.

19 **Is it better to believe falsehoods or reject truths?** Michael Shermer, editor of *Skeptic* magazine, calls these type I and type II errors, respectively. He argues that in making causal associations we are always faced with a choice between minimizing the statistical chance of making one error or the other (*Why People Believe Weird Things: Pseudoscience, Superstition, and Other Confusions of Our Time* [New York: W. H. Freeman & Co., 1998]).

19 **we have evolved a tendency to believe in the reality of the patterns that we perceive:** Foster and Kokko ("The Evolution of Superstitious and Superstition-Like Behavior") point out that "as long as the cost of type II errors [rejecting a truth] is high enough, natural selection can favour strategies that frequently make type I errors and generate superstitions." In other words, as they demonstrate with mathematical rigor, superstitions are adaptive.

19 **"Baloney Detection Network":** Michael Shermer, "Patternicity: Finding Meaningful Patterns in Meaningless Noise," *Scientific American* (December 2008): 24. The philosopher Bertrand Russell once advanced a "wildly paradoxical and subversive" doctrine that "it is undesirable to believe a proposition when there is no reason whatsoever for supposing it to be true" (*Skeptical Essays* [London: George Allen & Unwin, 1977], 11). This doctrine is a cornerstone of the scientific method, whose main object is to check beliefs against reality. I wonder what Russell would have made of Shermer's evolutionary argument?

19 **basic philosophical fallacies:** For an excellent list, see "Fallacies," Internet Encyclopedia of Philosophy, http://www.iep.utm.edu/fallacy/.

19 **"the post hoc fallacy" . . . was first described by . . . Aristotle:** According to *The Oxford Companion to Philosophy*, edited by Ted Honderich (New York: Oxford University Press, 1995), the phrase "post hoc" ergo propter hoc appears to derive from Aristotle's *Rhetoric.* The post hoc fallacy and its consequences are well described in "Fallacy: Post Hoc," The Nizkor Project, http://www.nizkor.org/features/fallacies/post-hoc.html.

20 **volatility in the housing market:** John Calverley, *When Bubbles Burst: Surviving the Financial Fallout* (London: Nicholas Brealey Publishing, 2009).

20 an increasing tendency to believe in the advice of experts: Elaine Scoggins, "Five Warning Signs That You're Caught Up in a Market Bubble," *Wall Street Journal*, October 30, 2009, http://www.marketwatch.com/story/five-signs-youre-caught-in-a-market-bubble-2009-10-30.

20 Cherry-picking: Cherry-picking is also known as "the fallacy of suppressed evidence" ("Fallacies," Internet Encyclopedia of Philosophy, http://www.iep.utm.edu/fallacy/#Cherry-Picking).

20 *Literary Digest*: On the poll in question, see "The Gallup Organization," http://www.answers.com/topic/the-gallup-organization.

21 modern-day data-miners: Data-mining is now big business. Other uses in addition to those listed include market analysis, health screening, and community planning. For a basic description of data-mining practices, see my book *The Perfect Swarm* (New York: Basic Books, 2009). For more advanced information, see Yi Peng, Gang Kou, Yong Shi, and Zhengxin Chen, "A Descriptive Framework for the Field of Data Mining and Knowledge Discovery," *International Journal of Information Technology and Decision Making* 7 (2008): 639–682; and Jiawei Han and Micheline Kamber, *Data Mining: Concepts and Techniques*, 2nd ed. (San Francisco: Morgan Kaufmann, 2006).

21 countries where people spend less time eating have higher economic growth rates: Floyd Norris, "Eat Quickly, for the Economy's Sake," *New York Times*, May 8, 2009, http://www.nytimes.com/2009/05/09/business/09charts.html?_r=1. For an excellent dissection of this piece of nonsense, see Gordon Linoff, "Not Enough Data," May 10, 2009, http://blog.data-miners.com/2009_05_01_archive.html.

22 the "false positive" . . . is becoming an increasingly serious issue: Ben Goldacre, "Datamining for Terrorists Would Be Lovely if It Worked," *The Guardian*, February 28, 2009, http://www.badscience.net/2009/02/datamining-would-be-lovely-if-it-worked/.

23 "History never repeats itself in exactly the same way": Makridakis and Taleb, "Decision Making and Planning Under Low Levels of Predictability."

23 "Black Swans": Nassim Taleb, *The Black Swan: The Impact of the Highly Improbable*, 2nd ed. (New York: Random House, 2010).

23 a classical philosophical example: This example first appeared in print by Karl Popper, *The Logic of Scientific Discovery* (London: Hutchinson & Co., 1959), 3.

23 Willem de Vlamingh . . . became the first European to see a *black* swan: R. H. Major, *Early Voyages to Terra Australis* (London: Hakluyt Society, 1859), available as a Project Gutenberg e-book at: http://gutenberg.net.au/ebooks06/0600361h.html. According to my history teacher at school, de Vlamingh was so surprised that he refused to believe that the birds really *were* swans. Like many another plausible historical anecdote, this one has no basis in fact. The hardheaded Dutch sea captain expressed no surprise when he wrote about the swans in his journal and merely took a couple of young ones (which later died) on board to show the folks back home in Batavia.

24 when a shallow lake "flips" from a clear to a turbid state: Marten Scheffer and Egbert H. van Nes, "Shallow Lakes Theory Revisited: Various Alternative Regimes Driven by Climate, Nutrients, Depth, and Lake Size," *Hydrobiologia* 584 (2007): 455–466.

CHAPTER 3: GALILEO'S HELL

25 **"Abandon hope, all ye who enter here":** Dante's "Inferno" is the first part of the *Divine Comedy*, a medieval allegorical poem about the soul's journey toward God. All of the action takes place from the night before Good Friday to the Wednesday after Easter in the year 1300.

The poem is divided into three parts; the second and third parts are called "Purgatorio" and "Paradiso," respectively. Dante's description of the geography of Hell comes in the first part, where the Roman poet Virgil guides him into its depths. Virgil is also his guide through Purgatory, while his "ideal woman," Beatrice, guides him through Heaven in the third and final part. The full text of the poem can be found as a Project Gutenberg e-book at: http://www.gutenberg.org/files/8800/8800.txt.

25 **"Heaven for climate, and Hell for society":** This often misquoted phrase comes from a political speech entitled "Tammany and Croker" that Mark Twain made to the "Order of Acorns" in 1901 (available as a Project Gutenberg e-book at: http://www.gutenberg.org/files/3188/3188.txt). The full quote is: "The election makes me think of a story of a man who was dying. He had only two minutes to live, so he sent for a clergyman and asked him, 'Where is the best place to go to?' He was undecided about it. So the minister told him that each place had its advantages—heaven for climate, and hell for society."

The Order of Acorns was organized around 1901 in support of New York mayoral candidate Seth Low and in opposition to the Tammany Hall political machine that had influenced New York City politics for decades. The Order still exists at the time of writing, with the "Chief Oak" being the deputy fire commissioner of New York.

25 **one of its cardinals asked . . . Galileo:** The person who seems most likely to have invited Galileo to perform his calculations and describe the results to the Florentine Academy was the newly appointed Cardinal Francesco del Monte; see Mark A. Peterson, "Galileo's Discovery of the Scaling Laws" (*American Journal of Physics* 70 [June 2002]: 575–580, http://scitation.aip.org/getabs/servlet/GetabsServlet?prog=normal&id=AJPIAS000070000006000575000001&idtype=cvips&gifs=yes), which gives an excellent, academically rigorous historical account of the evolution of Galileo's ideas from his lectures at the Florentine Academy to *Dialogues Concerning Two New Sciences*, written fifty years later.

25 **the twenty-four-year-old Galileo:** Galileo had established his reputation by calculating the center of balance for objects with different complicated shapes. He also analyzed Archimedes' method for using a balance to compare the densities of different substances. (This is the idea that supposedly caused Archimedes to jump out of his bath and run down the street naked shouting "Eureka!") The result was a book called *La Bilancetta* (*The Little Balance*), which is very readable, even today, and whose publication seems to have been the catalyst for Galileo's invitation.

For an outline of Galileo's life, work, and attempts to undermine the establishment, see my *Weighing the Soul: Scientific Discovery from the Brilliant to the Bizarre* (New York: Arcade, 2004), which also gives details of *The Little Balance* and the story of Archimedes in his bath.

26 **Hell must be shaped like an ice cream cone:** Galileo's exact words were: "It is in the shape of a concave surface which we call conical, its vertex is at the center of the world, and the base is against the surface of the earth."

26 **the vaulted roof . . . [spanned] a diameter equal to the radius of the earth:** Galileo used two very neat geometric arguments. The first came from the fact that Dante describes Hell and Purgatory as being symmetrically disposed about the center of the earth, and hence of the same size and shape. The second came from Dante's view when he emerged from Purgatory to see that "Already the Sun was joined to the horizon / Whose meridian circle covers / Jerusalem with its highest point" ("Purgatorio" II, 1–3). From this description, Galileo made a brilliant geometrical inference: "The size and depth of the Inferno is as great as the radius of the earth, and its mouth, which is the circle turned about Jerusalem, has for its diameter an equal size, *because under the arc of the sixth part of the circle is a chord equal to the radius*" (my italics). By my calculation, this means that the circumference of Hell passes through Marseilles and Tashkent, which makes sense in the former case, at least in my experience.

26 **the radius of the earth, which [Galileo] took to be . . . 3,245 miles:** The German Mercator expert Wilhelm Krücken has kindly researched the question for me, and it appears that Galileo based his value on Mercator's lectures in cosmography in 1559 to 1562 at the grammar school (Gymnasium) of Duisburg, published by his son Bartholemäus in 1563 as *Breves in sphaeram* (fol. f DE). According to Mercator's calculations: 1 degree (of the earth's circumference) = "quindecim miliaria germanica communica" (15 "German miles" [mgc]). So the earth's circumference in these units is 360 × 15 = 5,400 mgc, and its corresponding radius is 859 mgc. Since 1 mgc = 3.778 miles, this gives a radius of 3,245 miles, which checks out.

26 **the prestigious Florentine Academy:** The Florentine Academy was set up by the Medici dynasty (which had just ascended to the Italian nobility in the previous generation) and had as one of its chief functions the glorification of the Medici (and Florence) in every intellectual arena.

26 **We still have the transcripts of those lectures:** Galileo gave these lectures in 1588. The full text, translated into English by Mark Peterson of Mount Holyoke College, can be found at http://www.mtholyoke.edu/courses/mpeterso/galileo/inferno.html.

27 **Here's Dante's description of Nimrod:** "Inferno" XXXI, 58–60.

27 **The bronze pinecone . . . now resides . . . inside the Vatican museum:** I am indebted to Mark Peterson for this information.

27 **88 feet tall:** Galileo actually used the Italian medieval unit of the *braccio*, a unit of length almost exclusively used for cloth and whose precise value depended on the city you were living in! In Pisa a *braccio* corresponded to 583.6 millimeters, 12 *crazie*(s) (http://www.sizes.com/units/braccio.htm)—or, in modern units, about 2 feet.

28 **the world's currently tallest building:** That building, at 2,717 feet, is the Burj Khalifa in Dubai (http://www.burjkhalifa.ae/).

28 **[Galileo's calculation] . . . only worked, however, for a reason that he could not have been aware of:** The principle of universal gravitational attraction would be discovered later by Newton, who was born the year after Galileo died.

28 **Brunelleschi's famous . . . dome:** Many nice images are available online (see, for example, http://www.ciaoadriano.com/Europe%202007/Italy/04_8_Days/Images_8_Florence/900/Florence-Duomo-Nov07-RAW4827AR900.jpg). For a good account of the dome's design and construction, see Ross King, *Brunelleschi's Dome* (London: Vintage, 2008).

28 **the dimensions of Brunelleschi's famous . . . dome:** Jacques Heyman, *The Stone Skeleton: Structural Engineering of Masonry Architecture* (Cambridge: Cambridge University Press, 1995).

29 ***Dialogues Concerning Two New Sciences*:** This work was published in Amsterdam by Elsevier in 1638. The full text is available in English translation at: http://www.phys.virginia.edu/classes/109N/tns_draft/index.html.

29 ***Dialogues* . . . even [described] some scientific party tricks:** I describe one of these in *Weighing the Soul*, 206–207.

29 **Galileo's famous "square-cube law":** The consequences of this law for the size of bones and bodies are described in nontechnical detail at: http://www.dinosaurtheory.com/scaling.html. The law is, of course, a first approximation and an oversimplification (for example, bones are not perfect cylinders, and their material composition is not uniform). Scaling the strength of real bones is the subject of ongoing research; see, for example, Michael Doube et al., "Three-Dimensional Geometric Analysis of Felid Limb Bone Allometry," *PLoS ONE* 4(3, March 9, 2009): e4742, http://www.plosone.org/article/info:doi%2F10.1371%2Fjournal.pone.0004742.

29 **"whose natural length has been increased three times":** *Dialogue Concerning Two New Sciences*, translated by Henry Crew and Alfonso de Salvio (Macmillan, 1914), available through University of Virginia Electronic Text Center.

30 **it wasn't until the 1890s that the correct scaling law emerged:** Matthys Levy and Mario Salvadori, *Why Buildings Fall Down* (New York: W. W. Norton & Co., 2002), 37.

30 **the thickness has to increase with the *square* of the span:** Mario Salvadori and Matthys Levy, *Structural Design in Architecture* (New York: Prentice-Hall, 1967). Levy offers a nonquantitative example that the Romans apparently knew about. "A slab of stone can span 1m with a reasonable thickness but at 10m, the thicker slab collapses of its own weight" (personal communication, e-mail, August 7, 2010).

30 **a dome [of reinforced concrete] spanning 147 feet need only be 8 inches . . . thick:** Levy and Salvadori, *Why Buildings Fall Down*, 37. These authors describe an ingenious way of building such domes that was invented by an Italian architect called Bini (first name Dante!). It consists of laying an uninflated plastic balloon flat, placing reinforcing bars on it, and pouring concrete over them, then inflating the balloon with the concrete still wet and letting the concrete set. After this, the balloon is simply deflated and removed, and doors and windows are cut in the concrete.

31 **some 17 percent of medieval cathedrals collapsed soon after they were built:** Robert A. Scott, *The Gothic Enterprise: A Guide to Understanding the Medieval Cathedral* (Berkeley: University of California Press, 2003), 29–30.

31 **the . . . structure was . . . able to flex and either broke directly or vibrated itself to pieces:** Maury I. Wolfe and Robert Mark, "The Collapse of the Vaults of Beauvais

Cathedral in 1284," *Speculum* 51 (1976): 462–476; Philippe Bonnet-Laborderie, *Découvrir la Cathedrale Saint-Pierre de Beauvais* (2000).

CHAPTER 4: THE STRESS OF IT ALL

35 **"Beautiful Railway Bridge of the Silv'ry Tay!":** "The Tay Bridge Disaster" is the middle poem of a truly execrable set of three. The first ("Beautiful Railway Bridge of the Silvery Tay!") celebrates the original construction of the bridge in May 1879. The second mourns its collapse a scant nineteen month later, while the third, entitled "An Address to the New Tay Bridge," celebrates the completion of its successor in July 1887.

38 **"The Great McGonagall":** William McGonagall was a poorly educated hand-loom weaver who "discovered his discordant muse" in 1877 according to the website McGonagallOnline (which found this information in his autobiography, "Dame Fortune has been very kind to me by endowing me with the genius of poetry"). He sold his poems in printed broadsheets on the streets of Dundee, became locally famous through many articles in *Dundee* magazine, and developed a reputation as an overly persistent orator and performer in the pubs and theaters of his hometown. His international fame was assured after he was discovered by *Punch* magazine, which published many of his poems. For more on this infamous poet, see McGonagallOnline, http://www.mcgonagall-online.org.uk/. My favorite McGonagall quote (from his autobiography) is "the first man who threw peas at me was a publican."

35 **the collapse of the recently constructed Tay Railway Bridge:** The facts about the Tay Bridge disaster are described by Peter R. Lewis in *Beautiful Railway Bridge of the Silvery Tay: Reinvestigating the Tay Bridge Disaster of 1879* (Letchworth, U.K.: Tempus, 2004). See also the BBC/Open University forensic engineering website, Open2.NET (http://www.open2.net/forensic_engineering/), which presents a hands-on interactive analysis of the disaster. A good potted history is provided in "Tay Bridge and Associated Lines (North British Railway)," RAILSCOT, http://www.railbrit.co.uk/Tay_Bridge_and_associated_lines/frame.htm.

36 **"the insufficiency of the cross bracing . . .":** *Report of the Court of Inquiry and Report of Mr. Rothery upon the Circumstances attending the Fall of a Portion of the Tay Bridge on the 28th December 1879*, available at The Railways Archive, http://www.railwaysarchive.co.uk/docSummary.php?docID=107. The report castigated the engineer Thomas Bouch for his poor design and paid little heed to his defense that the wind load was far higher than he had been told was possible (by the Astronomer Royal, no less!). For more on the disaster, see V. Ryan, "The Tay Bridge Disaster," http://www.technologystudent.com/struct1/taybrd1.htm.

36 **the beautiful flying buttresses:** Flying buttresses are simply those that have a span passing through the air before they contact solid supports buried in the ground.

36 **The [Eiffel Tower] . . . flexes by two to three inches:** For a few statistics on the Eiffel Tower, see the Eiffel Tower website, http://www.tour-eiffel.fr/teiffel/uk/documentation/structure/page/chiffres.html.

36 **Guy de Maupassant . . . had lunch every day in the tower's restaurant:** Jill Jonnes, *Eiffel's Tower: And the World's Fair Where Buffalo Bill Beguiled Paris, the Artists*

Quarreled, and Thomas Edison Became a Count (New York: Viking Adult, 2009), 163–164.

37 **When the Tay Bridge was replaced:** The new bridge was completed in July 1887.

37 **"*Beautiful new railway bridge*":** There are many images online; for example, see the image at Mullys Webs.com, http://mullys.webs.com/nlfacts.htm.

38 **the Scottish Forth Rail Bridge . . . built by the British engineer Benjamin Baker:** See "The Forth Rail Bridge," Forth Bridges, Visitors Centre Trust, http://www.forthbridges.org.uk/railbridgemain.htm.

38 **The original contractor was Thomas Bouch:** It was particularly ironic that Bouch had called Baker as a witness in his defense after the Tay Bridge disaster. Bouch's main defense was that the wind load had been far higher than he had been advised was possible. To Bouch's dismay, Baker pulled the rug out from under him with calculations that proved the stresses caused by the wind load to be high, but not outside the bounds of normality.

38 **"Reality is the leading cause of stress!"** Script by Jane Wagner.

39 **the importance of wind-generated stresses:** See "Geometry and Materials," a nice analysis of the Eiffel Tower from the Whiting School of Civil Engineering at Johns Hopkins University at: http://www.ce.jhu.edu/perspectives/studies/Eiffel%20Tower%20Files/ET_Geometry.htm.

39 **built a wind tunnel at the base of his tower:** La Tour Eiffel, "Gustave Eiffel," http://www.tour-eiffel.fr/teiffel/uk/documentation/dossiers/page/gustave_eiffel.html.

39 **loaded airplane suspended by its wingtips on huge "jacks" designed for the purpose:** For an eye-popping video of how far the wings bend when the plane is deliberately dropped, see "In Big Test, Boeing 787 Wing Bends, Doesn't Break," Telstar Logistics, March 29, 2010, http://telstarlogistics.typepad.com/telstarlogistics/2010/03/in-extreme-test-boeing-787-wing-bends-but-does-not-break.html.

39 **Cardiologists use treadmill stress tests:** Myrvin H. Ellestad, William Allen, Maurice C. K. Wan, and George L. Kemp, "Maximal Treadmill Stress Testing for Cardiovascular Evaluation," *Circulation* 39 (1969): 517–522.

39 **a psycho-social stress test:** Clemens Kirschbaum, Karl-Martin Pirke, and Dirk H. Hellhammer, "The 'Trier Social Stress Test'—A Tool for Investigating Psychobiological Stress Responses in a Laboratory Setting," *Neuropsychobiology* 28 (1993): 76–81.

39 **major banks . . . are now undergoing stress tests:** Margaret Popper, "Bank Stress Test—How It Works," Bloomberg News, April 23, 2009, http://www.youtube.com/watch?v=kCAlWBHB0XA.

40 **The precise physical definition [of stress]:** According to the reviewer Ian Sneddon (*Bulletin of the American Mathematical Society* 3 [1980]: 870–878):

> Cauchy's memoir on the subject was communicated to the Paris Academy in September 1822 but it was not published. An abstract appeared in the *Bulletin des Sciences à la Société Philomathique*, 1823 and the contents were fully described in three articles in Cauchy's *Exercises de Mathématique* (1827, 1828). The third of these, published in the volume for 1828, entitled "Sur les équations qui experiment les conditions d'équilibre ou les lois de

mouvement intérieur d'un corps solide" established the correct equations to model the behavior of an isotropic elastic body.

40 **stress is the *force* . . . divided by the *area*:** See J. E. Gordon, *The New Science of Strong Materials or Why You Don't Fall Through the Floor* (Harmondsworth, U.K.: Penguin Books, 1968), 34, for an excellent elementary discussion. A highly recommended read!

40 **It is tempting to think that Cauchy was inspired:** It is probably wishful thinking to think his family's experience with the guillotine would have had this effect on Cauchy. By all accounts, he was a rather self-centered man.

40 **A lot of confusion could be avoided if we used the term "strain":** The confusion arose because the Hungarian-Canadian endocrinologist Hans Selye had only a limited command of English when he introduced the concept into biology in 1937. He later admitted that, had his knowledge of English been more precise, he would have gone down in history as the father of the "strain" concept rather than the "stress" concept.

Selye was looking for a word to describe his discovery that animal bodies have a very similar set of responses to a wide variety of "nocuous agents," such as cold, pain, illness, or drugs. Not knowing that the word already had a precise physical definition, he decided to call this set of responses "stress," which he defined as "the non-specific response of the body to any demand for change." He first attempted to do so in his seminal article "A Syndrome Produced by Diverse Nocuous Agents" (*Nature* 138 [1936]: 32), but at an editor's suggestion, he replaced the term with "general alarm reaction." His first actual use of the word "stress" in print was in an article entitled "Studies on Adaptation" (*Endocrinology* 21 [1937]: 169–188).

The confusion between the physical and biological meanings of stress has been exacerbated by the fact that it is now a popular buzzword, with such a wide variety of meanings that it has become essentially meaningless. That confusion was spelled out by the psychoanalyst Robert Dato in "The Law of Stress" (*International Journal of Stress Management* 3 [1996]: 181–182): "Physical scientists use 'stress' to indicate a force, pressure, or stimulus, whereas biological scientists use 'stress' to indicate a change or response. The opposite use of the term by these two groups is unfortunate because it confuses the meaning of the concept for the scientific community as well as the general public."

The meaning of stress varies even among professional psychologists. Some use it to mean a response, some see it as being applied from the outside, and some even use it interchangeably to mean both. Selye himself was guilty of ignoring his own definition on occasion and used the word interchangeably to mean a cause in some instances and an effect in others. An examination of verbatim citations from his own writings led to the conclusion that "stress, in addition to being itself, was also the cause of itself, and the result of itself"; see Paul J. Rosch, "Reminiscences of Hans Selye, and the Birth of 'Stress,'" American Institute of Stress, http://www.stress.org/hans.htm.

For a history of the subsequent use of the concept of stress in psychology, see American Institute of Stress, "Stress, Definition of Stress, Stressor, What Is Stress? Eustress?" http://www.stress.org/topic-definition-stress.htm; and Rosch, "Reminiscences of Hans Selye."

"If you were to ask a dozen people to define stress," says the official website of the American Institute of Stress, "or explain what causes stress for them, or how stress affects them, you would likely get 12 different answers to each of these requests. The reason for this is that there is no definition of stress that everyone agrees on." The institute concludes that stress "is not a useful term for scientists because it is such a highly subjective phenomenon that it defies definition." Presumably the scientists referred to are mainly psychologists; physical scientists have no trouble with the term at all (American Institute of Stress, "Stress, Definition of Stress, Stressor").

41 **Galileo produced one of the first descriptions:** The effect that Galileo describes would vitiate my own solution to his problem with the roof of Hell, which was to support the roof with a series of columns, since materials are much stronger in compression than they are when they are being bent or twisted. The lower parts of my columns would have melted, however, under the temperatures that we now believe exist at the center of the earth. They would then have sunk, just as one of the marble columns did in his example, leaving areas of high local stress at the still-supported parts of the roof.

42 **The key insight [was] due to Cauchy:** J. E. Gordon, *Structures or Why Things Don't Fall Down* (Harmondsworth, U.K.: Penguin Books, 1978), 46.

42 **Leonardo da Vinci recognized this fact a hundred years before:** Professor Robert Ballarini from Case Western University points this out in "The Da Vinci–Euler–Bernoulli Beam Theory?," *Mechanical Engineering,* April 18, 2003, http://www.memagazine.org/contents/current/webonly/webex418.html; see also Ladislao Reti, ed., *The Unknown Leonardo* (New York: McGraw-Hill, 1974).

43 **using fine glass springs to measure the tiny forces:** V. M. Bowers, L. R. Fisher, G. W. Francis, and K. L. Williams, "A Micromechanical Technique for Monitoring Cell-Substrate Adhesiveness: Measurements of the Strength of Red Blood Adhesion to Glass and Polymer Test Surfaces," *Journal of Biomedical Materials Research* 23 (1989): 1453–1473.

44 **the English scientist Robert Hooke:** Hooke's life and achievements are splendidly recorded in Stephen Inwood's *The Man Who Knew Too Much* (London: Macmillan, 2002). Among his many discoveries were the spring-regulated watch, the first practical Gregorian reflecting telescope, a calculating machine, and the uses and effects of cannabis. According to Hooke, "The science of Nature has been already too long made only a work of the brain and the fancy: It is now high time that it should return to the plainness and soundness of observations on material and obvious things" (*Micrographia,* 1665). For a description of Hooke's employment at the Royal Society, see my *How to Dunk a Doughnut* (New York: Arcade, 2002), 3–4.

44 **[Hooke's] law of elasticity:** For an excellent (though somewhat technical) history of elasticity, see Ian Sneddon's book review of V. D. Kupraze et al., "Three-Dimensional Problems of the Mathematical Theory of Elasticity and Thermoelasticity," *Bulletin of the American Mathematical Society* 3 (1980): 870–878, http://projecteuclid.org/euclid.bams/1183547551.

44 **[Hooke] published [the law of elasticity] as the anagram: *ceiiinosssttuu:*** The anagram appeared in a book about methods for observing eclipses, *A description of*

helioscopes, and some other instruments (London: John Martin, 1676)! To fill up the white space at the end of the last page, he added: "A decimate of the centesme of the inventions I intend to publish," followed by a list of ten that included "3. The true Theory of Ela ticity or Springine s, and a particular Explication thereof in everal Subjects in which it is to be found: And the way of computing the velocity of Bodies moved by them. ceiiinosssttuu." "Some other instruments" included a way to regulate pendulum clocks, a method for constructing arches, and inventions in optics, hydraulics, and mechanical engineering. Two years later he published the solution to the anagram in *De Potentia Restitutiva, or of Spring Explaining the Power of Springing Bodies* (London, 1678), 23. He never published the other 9,990 inventions that he had promised.

45 ***Course of Lectures on Natural Philosophy and the Mechanical Arts:*** This work can be read online at: http://www.archive.org/details/lecturescourseof02younrich.

45 **delivered at London's Royal Institution in 1807:** The Royal Institution is still very much alive and active. The lecture theater where Young spoke has been preserved as it was. I have even spoken there myself, although I am not sure whether my chosen subject ("The Science of Toffee Apples") would necessarily have met with Young's approval. He was a rather serious man.

47 **the shock wave from an exploding bomb generated tensile forces:** Even pile drivers can surprisingly generate damaging tensile forces. When concrete bridge supports are being driven into a seabed or a riverbed, the impact of the pile driver generates a compression wave that travels to the bottom of the pile, from which it is reflected back up. If the pile-driver hammer has bounced off before the return wave gets there, the return wave can create momentary but intense tension in the concrete at the top of the pile that is sufficient to shatter it. Pile drivers have to be carefully designed and operated to avoid this contingency by "damping" the hammer blow so that the hammer is still in contact with the top of the pile to absorb the force when the return wave hits. Apparently this observation was an inspiration for Barnes Wallis, the inventor of the "bouncing bomb" that did so much damage to the German dams. On the "Dambuster" raids, see http://www.dambusters.org.uk/.

48 **One of the most unusual . . . ways [to measure stress] was to measure the shape of soap bubbles:** G. I. Taylor and A. A. Griffith, "Use of Soap Films in Solving Torsion Problems" and "The Application of Soap Films to the Torsion and Flexure of Hollow Shafts," reprinted in *The Scientific Papers of Sir Geoffrey Ingram Taylor*, http://books.google.com.au/books?hl=en&lr=&id=U8w7AAAAIAAJ&oi=fnd&pg=PA1&dq=The+application+of+soap+films+to+the+torsion+and+flexure+of++hollow+shafts&ots=h9EqSKHDPt&sig=a1kZQzkGoG9ic1TRwG6bMBwgwDk#v=onepage&q=The%20application%20of%20soap%20films%20to%20the%20torsion%20and%20flexure%20of%20%20hollow%20shafts&f=false. These two papers are really worth a read by anyone interested in how simple experiments can produce profoundly interesting results.

48 **Griffith showed that stress becomes *concentrated* at the tips of cracks:** This demonstration was made in a paper that Griffith read to the Royal Society in February 1920 and that appeared in print as "The Phenomena of Rupture and Flow in Solids,"

Philosophical Transactions of the Royal Society of London A 221 (1921): 163–198, http://www.jstor.org/stable/91192.

49 **the Liberty ship *Schenectady*. . . suddenly split in half:** See, for example, Norbert J. Delatte, *Beyond Failure: Forensic Case Studies for Civil Engineers* (Reston, Va.: American Society of Civil Engineers, 2008), 311.

49 **glaciers when they "calve":** For an excellent example, see "Glacier Calving," YouTube, September 18, 2006, http://www.youtube.com/watch?v=bYH2Df-evNs. Look for the critical crack!

50 **These composites act as *crack-stoppers*:** One example is the chocolate-coated cookie, which is less likely to be broken during transport than its uncoated cousin.

51 **Some 20 percent of ships now at sea have cracks somewhere near this length, and occasionally one does [split in two]:** On August 19, 2007, for example, an oil tanker off the coast of Australia split in two, dumping twenty thousand tons of crude oil into the ocean. For a wonderfully deadpan takeoff on the Australian politicians' response and attempted justification, see "Front Fell Off," YouTube, February 4, 2007, http://www.youtube.com/watch?v=WcU4t6zRAKg.

51 **the commercially available *dye penetrant detector test*:** See, for example, Les Bengtson, "Crack Inspection for the Hobbyist" (2003), http://www.custompistols.com/cars/articles/crack_inspection.htm. An ex-girlfriend of mine who was a very keen cyclist tried this test on the pedal cranks of her bicycle and was sufficiently alarmed by the results to have them replaced immediately. Confident of their strength, she applied the maximum force as she tried to race a car from the lights. The cranks held—but the stress split her kneecap!

CHAPTER 5: RUNAWAY DISASTER

53 ***Stop the World—I Want to Get Off!*:** This musical by Leslie Bricusse and Anthony Newley was first produced in London's Queen's Theatre in 1961. The title is said to have been derived from a piece of wall graffiti.

53 **the 1980 *Blues Brothers* film:** The original script is available at Blues Brothers Central, http://www.bluesbrotherscentral.com/movies/the-blues-brothers/script/.

54 **reaches a *point of no return* when it leaves the road:** There are many equivalent expressions taken from different areas of life and different eras in history: *crossing the Rubicon* (Julius Caesar); *the die is cast* (Julius Caesar, after crossing the Rubicon); *burning one's boats* (the Muslim commander *Tariq ibn Ziyad* on invading Spain in AD 711, also ironically the Spanish conquistador Hernán Cortés on invading the Aztec kingdom some seven hundred years later); and *fait accompli* (French). My favorite is the ancient Chinese expression *break the woks and sink the boats.* In the aircraft industry, the point of no return can also refer to the point in a flight where the aircraft can no longer get back to the airfield of departure (given by the *radius of action* formula). An aircraft flying through space might even reach the *event horizon*, which defines the point of no return for light being sucked into a black hole.

54 **the dialogue . . . brings the problem into focus:** The quote is from scenes 323-B and 324-B.

54 **Newton's . . . *three laws of motion*:** Isaac Newton, *Philosophiae Naturalis Principia Mathematica* (1687). For some good animations, see University of New South Wales, School of Physics, "Physclips," http://www.animations.physics.unsw.edu.au/mechanics/chapter5_Newton.html; and A-Level Physics Tutor, http://www.a-levelphysicstutor.com/yt-mech-newtons-laws.php. There are some obvious provisos for applying these laws to the movement of the Winnebago, the most important of which is the effect of friction, which always acts in the opposite direction to velocity.

55 **the floor would have been pushing back up with equal force:** To see this, just consider that the "Good Ol' Boys" might have been bouncing up and down, but they stayed in pretty much the same place, rather than accelerating up into the sky or down into the ground. According to Newton's second law, if there is no net vertical acceleration, then there is no net vertical force, which means that *something*—the reaction force from the floor in this case—must be exactly balancing the downward force of their weight.

55 **Newton's laws govern the movement of every physical object:** To avoid incurring the wrath of my physicist colleagues, I should point out that this is strictly true only when the laws are updated to account for the relativistic effect that mass increases with speed.

55 **cartoon characters . . . obey a set of laws that are entirely their own:** Mark O'Donnell, "The Laws of Cartoon Motion," in *Elementary Education: An Easy Alternative to Actual Learning* (New York: Random House, 1985), also available at: http://www.rahul.net/figmo/Archives/toon-physics.html.

55 **Wile E. Coyote, the Newton of cartoon physics:** See, for example, the following YouTube clips: Looney Tunes, "Coyote Fall," http://www.youtube.com/watch?v=_d8ROhH3_vs; "Wile E. Coyote—Way to Fall Down," http://www.youtube.com/watch?v=RrZiFzEPz-I&feature=related; "Wile E. Coyote: Fail vs. Epic Fail," http://www.youtube.com/watch?v=Hd755bbc8uw&feature=related.

55 **the parallel with the behavior of many real-life participants in bull markets:** Shawn Andrew, "The Cartoon Law," *Reflexivity Capital Group Investment Framework* (2007), www.refcapgroup.com/pdf/strategy-rcg.pdf.

57 **Running, Jumping, and Standing Still:** The subhead is borrowed from the anarchic comedy film *The Running, Jumping, and Standing Still Film*, written by Spike Milligan, Peter Sellers, and Richard Lester and produced by Peter Sellers in 1959 as his first venture into filmmaking after the long-running radio show *The Goon Show*. This famous film is just eleven minutes long and cost around $150 to make! See The Telegoons, "Running Jumping & Standing Still . . . ," http://telegoons.org/history_4_running_jumping.htm.

57 **a nice little quirk in the algebra:** Newton's law of universal gravitation says that the force of gravitational attraction Fg between two objects of mass m and M whose centers of mass are a distance d apart is $Fg = G(m \times M)/d^2$, where G is the "universal gravitational constant." If we call M the mass of the earth, and m the mass of a much smaller object falling toward it, then the acceleration g of the object is $g = Fg/m$. Put the two equations together and m cancels out, leaving $g = (G \times M)/d^2$.

G, M, and d all have known, constant values. Just plugging them in (or measuring g directly!) gives us $g = 9.83$ meters/second^2, or 32 feet/second^2.

57 Galileo [dropping] cannonballs from a tower: As I show in *Weighing the Soul* (New York: Arcade, 2004), 23–24, the tower involved in this experiment was most unlikely to have been the Leaning Tower of Pisa. It isn't high enough, for one thing. Galileo said that he dropped his cannonballs 300 feet, but the Leaning Tower is only 150 feet high.

58 hundreds of people die . . . from falling out of bed: The statistic is based on National Safety Council estimates, available at "The Most Common Causes of Death Due to Injury in the United States," http://danger.mongabay.com/injury_death.htm.

58 the famous annual cheese-rolling contest: See "Gloucestershire Cheese Rolling," SoGlos.com, http://www.soglos.com/sport-outdoor/27837/Gloucestershire-Cheese-Rolling-2009.

58 Isaac Newton . . . [tried] to decipher the prophecies of the Bible: See Newton's *Observations upon the Prophecies of Daniel, and the Apocalypse of St. John*, published posthumously in 1733 (available as a Project Gutenberg e-book at: http://www.gutenberg.org/etext/16878). Newton argued that biblical prophecies could only be truly understood once the events that they foretold had come to pass. "The folly of interpreters," he said,

> has been to foretell times and things by this prophecy, as if God designed to make them prophets. By this rashness they have not only exposed themselves, but brought the prophecy also into contempt. The design of God was much otherwise. He gave this and the prophecies of the Old Testament, not to gratify men's curiosities by enabling them to foreknow things, but that after they were fulfilled they might be interpreted by the event, and his own providence, not the interpreters', be then manifested thereby to the world. For the event of things predicted many ages before will then be a convincing argument that the world is governed by Providence.

Newton's praiseworthily cautious approach has not been adopted by some present-day prophets—such as Pastor David Wilkerson of New York, who in March 2009 predicted an "earth-shattering calamity" in New York ("David Wilkerson Again Predicts Catastrophe," ReligionNewsBlog, March 11, 2009, http://www.religionnewsblog.com/23331/david-wilkerson-prophecy). If it happens, this book will not be published.

60 yet another customer for the waiting ambulance men: This is when participants discover the meaning of the event's full name: "Cheese Rolling and Wake."

61 Bedouin tribesmen call this effect the *camel's nose*: Mario J. Rizzo and Douglas Glen Whitman, "The Camel's Nose in the Tent: Rules, Theories, and Slippery Slopes," *UCLA Law Review* 51 (2003): 539–592.

61 as in the famous Honda TV advertisement: See "Honda Domino Advert," YouTube, May 7, 2007, http://www.youtube.com/watch?v=cQJPYgl5aoY.

61 Domino toppling is now an art form: See, for example, an artistic display at the Brattleboro Museum and Art Center in Vermont, "Domino Toppling 2: Brattleboogaloo," YouTube, March 10, 2009, http://www.youtube.com/watch?v=6mzqRouE_hs.

61 **Speca . . . tried for a new world record:** "College Senior Bob Speca Finds It's No Pushover to Set a World Record for the Domino Effect," *People*, June 26, 1978, http://www.people.com/people/archive/article/0,,20071149,00.html.

61 **the record for domino toppling:** This is the record for a group of people setting the dominoes up. The individual record, currently held by Ma Li Hua of China, now stands at 303,621; see "Domino Toppling Records," http://www.recordholders.org/en/records/domino-toppling.html.

61 **As wave after wave of dominoes fell:** See "Domino Toppling World Record," http://noolmusic.com/myspace_videos/domino_toppling_world_record.php.

62 **a seventeen-mile-long row of bricks:** See "Aktion 'Falling Stones' anlässlich der Wesertunnel-Eröffnung," January 20, 2004, http://www.asc-bremerhaven.de/index.php?option=com_content&view=article&id=28%3A17012004-aktion-qfalling-stonesq-anlich-der-wesertunnel-erung&catid=23%3Aaktiv&Itemid=33.

62 **Theory suggests that the optimum spacing:** Stan Wagon, William Briggs, and Stephen Becker, "The Dynamics of Falling Dominoes," *UMAP Journal* 26 (2005): 35–48.

62 **the rate of collapse is expected to be constant:** W. J. Stronge and D. Shu, "The Domino Effect: Successive Destabilization by Cooperative Neighbors," *Proceedings of the Royal Society of London A* 418 (1988): 155–163; Wagon et al., "The Dynamics of Falling Dominoes."

62 **The domino effect . . . has . . . been implicated in computer password theft:** Blake Ives, Kenneth R. Walsh, and Helmut Schneider, "The Domino Effect of Password Reuse," *Communications of the Association for Computing Machinery* 47 (2004): 75–78.

63 **A similar "gap-creating" approach is now being used to anticipate and avoid major industrial accidents:** Faisal I. Khan and S. A. Abbasi, "An Assessment of the Likelihood of Occurrence, and the Damage Potential of Domino Effect (Chain of Accidents) in a Typical Cluster of Industries," *Journal of Loss Prevention in the Process Industries* 14 (2001): 283–306; J. R. B. Alencar, R. A. P. Barbosa, and M. B. de Souza Jr., "Evaluation of Accidents with Domino Effect in LPG Storage Areas," *Thermal Engineering* 4 (2005): 8–12.

64 **the penultimate scene in the film *Zorba the Greek*:** The 1964 film was based on the 1946 novel of the same name by the Greek author Nikos Kazantzakis. The scene with the ramshackle ramp collapsing occurs in the film but not in the book.

64 **The progression of the [Interstate 35 West bridge] collapse was captured on a security camera:** See "35W Bridge Collapse," YouTube, August 2, 2007, http://www.youtube.com/watch?v=osocGiofdvc.

64 **The primary cause was . . . the fracture of undersized gusset plates:** Reggie Holt and Joseph Hartmann, "Adequacy of the U10 & L11 Gusset Plate Designs for the Minnesota Bridge No. 9340 (I-35W over the Mississippi River)," Federal Highway Administration Turner-Fairbank Highway Research Center report (interim report), January 11, 2008.

64 **Ronan Point Towers [collapse]:** "The Ronan Point Apartment Tower Case," summarized from Cynthia Rouse and Norbert Delatte, *Lessons from the Progressive Collapse of the Ronan Point Apartment Tower*, proceedings of the Third ASCE Forensics Congress, San Diego, California, October 19–21, 2003.

The most chilling example of a progressive collapse is, of course, that of the World Trade Center towers in the wake of the attacks of September 11, 2001. Its causes are still being argued over, but there seems little doubt that collapsing upper floors in each building initiated the collapse of floors below them as they fell in a progressive sequence that eventually felled the entire building. See Zden k P. Bažant, Jia-Liang Le, Frank R. Greening, and David B. Benson, "Closure to 'What Did and Did Not Cause Collapse of World Trade Center Twin Towers in New York?'" *Journal of Engineering Mechanics* 136 (2010): 934–935, http://scitation.aip.org/getabs/servlet/GetabsServlet?prog=normal&id=JENMDT000136000007000934000001&idtype=cvips&gifs=yes.

65 **Domino effects have been implicated in the nationalization of oil industries:** Stephen J. Kobrin, "Diffusion as an Explanation of Oil Nationalization: Or the Domino Effect Rides Again," *Journal of Conflict Resolution* 29 (1985): 3–32; **the melting of arctic glaciers:** Faezeh M. Nick et al., "Large-Scale Changes in Greenland Outlet Glacier Dynamics Triggered at the Terminus," *Nature Geoscience* 2 (2009): 110–114; **the blowdown of trees in pre-settlement Wisconsin forests:** Charles D. Canham and Orie L. Loucks, "Catastrophic Windthrow in the Presettlement Forests of Wisconsin," *Ecology* 65 (1984): 803–809; **menopausal symptoms:** E. M. Alder, L. A. Ross, and A. Gebbie, "Menopausal Symptoms and the Domino Effect," *Journal of Reproductive and Infant Psychology* 18 (2000): 75–78; **the progression of courtroom legal argument:** Rizzo and Whitman, "The Camel's Nose in the Tent."

66 **contagious spread of disease:** See, for example, E. C. Riley, G. Murphy, and R. L. Riley, "Airborne Spread of Measles in a Suburban Elementary School," *American Journal of Epidemiology* 107 (1978): 421–432.

66 **It is even possible to plan the . . . distribution of vaccines:** Edward Goldstein et al., "Distribution of Vaccine/Antivirals and the 'Least Spread Line' in a Stratified Population," *Interface* (*Journal of the Royal Society*) 7 (2010): 755–764, http://rsif.royalsocietypublishing.org/content/7/46/755.short.

66 **the key role played by *proximity*:** One of the designs for a nuclear bomb, for example, is called the "implosion assembly" (Carey Sublette, "Introduction to Nuclear Weapon Physics and Design," February 20, 1999, http://nuclearweaponarchive.org/Nwfaq/Nfaq2.html). It depends on the mass of fissile material being suddenly compressed so that the atoms are brought close enough together for a runaway chain reaction to take off.

67 **the contagious spread of laughter:** See, for example, the video of a baby laughing at "Skype Laughter Chain," YouTube, June 18, 2008, http://www.youtube.com/watch?v=p32OC97aNqc&annotation_id=annotation_693399&feature=iv. Can you stop laughing yourself in response?

67 **"Many years ago this was a thriving, happy planet":** Douglas Adams, *The Restaurant at the End of the Universe* (London: Pan Books, 1980), 61.

68 **Positive feedback can lead to many other critical transitions:** Some of these critical transitions do indeed have a positive outcome. One example is childbirth: an initial contraction releases the hormone oxytocin into the bloodstream, which stimulates further contractions, which release even more oxytocin, so that the oxytocin levels rise ever

higher and the contractions become more and more frequent until the child is born (for an excellent detailed account, in relatively simple language, not only of childbirth but also of the various other feedback processes that occur in the endocrine systems of our bodies, see "Endocrine System Information," http://www.besthealth.com/besthealth/bodyguide/reftext/html/endo_sys_fin.html.

Positive feedback has other biological uses. It is essential for the transmission of nerve signals, and our bodies use it in the polymerase chain reaction (see Molecular Station, "PCR Polymerase Chain Reaction" [video], http://www.molecularstation.com/science-videos/video/15/pcr-polymerase-chain-reaction/) to rapidly make multiple copies of DNA. (Forensic scientists use the same reaction to produce analyzable quantities of DNA from tiny samples.) Its physiological effects may also underlie our urge to learn and master certain skills.

CHAPTER 6: THE BALANCE OF NATURE AND THE NATURE OF BALANCE

71 **". . . a super-abundance of dreams":** Peter Ustinov, quoted in an interview for *The Independent* (United Kingdom), February 25, 1989.

71 **the "pumpkinseed":** The pumpkinseed is actually a sunfish, *Lepomis gibbosus* (see FishBase, http://www.fishbase.org/home.htm). The name probably derives from the shape, which resembles that of a pumpkin seed, and also from the dominant orange-yellow coloration. The seed would have to be a pretty big one, since the fishes can grow up to sixteen inches in length and over a pound in weight, although they normally reach only six to eight inches.

Stephen Forbes called this fish a "barbaric bream." The ichthyologist Mike Retzer, who is the collection manager for the Illinois Natural History Survey Fish Collection, concurs with me in believing that this must surely have been a local name, or maybe one invented by Forbes himself. No other writer has used it, and Forbes doesn't even mention it in the official *Fishes of Illinois*, which he edited in 1908.

71 **"an equilibrium of organic life":** This and other quotations by Forbes come from his paper "The Lake as a Microcosm," *Bulletin of the Scientific Association* (Peoria, Ill.) (1887): 77–87, http://people.wku.edu/charles.smith/biogeog/FORB1887.htm.

"The Lake as a Microcosm" was subsequently reprinted in the *Bulletin of the Illinois State Natural History Survey* (1925), where it finally became available to the wide audience that it deserved. Apparently the 1887 issue of the original *Bulletin* was the only one that the now-extinct Peoria Scientific Association ever published! A few copies of this rare publication still exist (G. T. Tonapi and V. A. Ozarkar, "A New Record of *Hydraena quadricollis Wollaston* [Coleoptera: Hydrophilidae] from India," *Coleopterists Bulletin* 23, no. 1 [March 1969]: 1–4, http://www.jstor.org/stable/3999265).

72 **Stephen Forbes—a self-taught scientist:** Forbes came from a pioneer family and had little formal education beyond the age of fourteen. Despite this handicap, he eventually became director of the Illinois State Laboratory of Natural History, chairman of the Department of Zoology at the University of Illinois, founder of the Illinois Biological Station, and chief of the Illinois Natural History Survey. He was elected to the National Academy of Sciences in 1918 and became president of the Ecological Society of America in 1921.

Quite a record for someone who was self-taught, and one that seems hardly possible these days.

A detailed biography and outline of Forbes's contribution to modern ecology are given in R. A. Croker, *Stephen Forbes and the Rise of American Ecology* (Washington, D.C.: Smithsonian Institute, 2001). His son, Ernest Browning Forbes, gave many interesting details of Forbes's poor background and personal life in his speech at his father's memorial service (Ernest Browning Forbes, *Memorial of the funeral services for Stephen Alfred Forbes, Ph.D., LL.D.: chief, state natural history survey, professor of entomology, emeritus, University of Illinois* [Urbana: University of Illinois Press, 1930?]). Among other things, Ernest Browning Forbes revealed his father's lifelong interest in the poet Robert Browning—the source of his middle name.

72 [Forbes's] pioneering studies of the northern Illinois lakes: The principal lakes concerned were Fox Lake, Nippersink Lake, Long Lake, Deep Lake, and Cedar Lake.

72 one of the first scientific explanations of what has become known as the *balance of nature*: Forbes, "The Lake as a Microcosm."

73 That honor goes to the ancient Greek historian Herodotus: Frank N. Egerton, "Changing Concepts of Balance of Nature," *Quarterly Review of Biology* 48 (1973): 322–350.

73 Forbes was a rationalist and agnostic: In 1923 he wrote: "I was, and still am, a rationalist and an agnostic, for whom what is known as faith is merely assumption." Sometime, however, between the time of this writing and his death, he crossed out the words "and still am" and replaced them with "as a younger man." According to his son (Forbes, *Memorial of the funeral services . . .*), the change was dictated by a growing hope that there was more to life than a scientist could know.

73 the fifth edition of his book *On the Origin of Species*: Charles Darwin, *On the Origin of Species by Means of Natural Selection, or the Preservation of Favoured Races in the Struggle for Life*, 5th ed. (London: John Murray, 1869): 91–92.

73 The phrase "survival of the fittest" was coined by . . . Spencer: Herbert Spencer, *Principles of Biology*, vol. 1 (1864), 444. Darwin first used the term in *The Variation of Animals and Plants under Domestication* (London: John Murray, 1868), where he acknowledged Spencer's coinage on page 6: "This preservation, during the battle for life, of varieties which possess any advantage in structure, constitution, or instinct, I have called Natural Selection; and Mr. Herbert Spencer has well expressed the same idea by the Survival of the Fittest. The term 'natural selection' is in some respects a bad one, as it seems to imply conscious choice; but this will be disregarded after a little familiarity" (see The Complete Work of Charles Darwin Online, http://darwin-online.org.uk/).

73 [the phrase "survival of the fittest"] would be open to a gross misinterpretation: Martin Shubik, "Does the Fittest Necessarily Survive?" in *Readings in Game Theory and Political Behavior*, edited by Martin Shubik (New York: Doubleday, 1954), 43–46, reprinted in Eric Rasmusen, ed., *Readings in Games and Information* (Malden, Mass.: Wiley-Blackwell, 2001), 105.

74 his hour-and-a-half-long talk: This is my own estimate, based on my personal speaking rate and the fact that the published version of Forbes's talk is nine thousand words long.

74 Many of Forbes's listeners . . . [with] similar Puritan upbringings . . . would have been taught, like him, to take a holistic view of nature: Nearly all early American ecologists had roots in churches that descended from the Puritan tradition. The historian Mark Stoll argues that this fact is highly significant ("Creating Ecology: Protestants and the Moral Community of Creation," in *Religion and the New Ecology: Environmental Responsibility in a World in Flux*, edited by David M. Lodge and Christopher Hamlyn [Notre Dame, Ind.: University of Notre Dame Press, 2006], 53–72).

74 One of the few people who never mastered [driving] was Stephen Forbes: Forbes, *Memorial of the funeral services. . . .*

76 One of the earliest examples: For a nice list from historical times to the present, see Frank L. Lewis, "A Brief History of Feedback Control," in *Applied Optimal Control and Estimation*, edited by Frank L. Lewis (New York: Prentice-Hall, 1992), http://www.theorem.net/theorem/lewis1.html.

76 the magnetic compass: An even more interesting, but less familiar, example is the "south-pointing chariot." Invented by the Chinese some two millennia ago, this chariot used a set of differential gears attached to the wheels so that a figure mounted on the front always pointed south, no matter which direction the chariot was going (see "Chinese South-Pointing Chariot," Science Museum [London], http://www.sciencemuseum.org.uk/objects/navigation/1952–275.aspx; Lu Jingyan, "Studies of the South-Pointing Chariot: Survey of the Past Eighty Years," in *Chinese Studies in the History and Science of Technology*, edited by Dainian Fan and Robert Sonné Cohen [Berlin: Springer, 1996], 267–278).

76 centrifugal flyball governor: James Patrick Muirhead, *The Life of James Watt: With Selections from His Correspondence* (1858; reprint, London: Nabu Press, 2010). For a nice video example of the governor in operation, see "Steam Engine Centrifugal Governor," YouTube, December 10, 2008, http://www.youtube.com/watch?v=EguLmX3o30w.

76 a water clock devised by the Greek inventor Ctesibius: This water clock is also called a *clepsydra* (J. S. McNown, "When Time Flowed: The Story of the Clepsydra," *La Houille Blanche* 5 [1976]: 347–353). It used a float to control the inflow of water through a valve; as the level of water fell, the valve opened and replenished the reservoir, thus keeping it at a constant level. This reservoir dripped water into a second, lower tank through a hole in the bottom. Because of the constant level in the upper tank, the lower tank filled at a constant rate and could thus act as a clock. Simple!

For a technical description and history of the clepsydra, see Silvio A. Bedini, "The Compartmented Cylindrical Clepsydra," *Technology and Culture* 3 (1962): 115–141, http://www.jstor.org/stable/3101437. For an animation, see "Ancient Greece: Animation: The Water Clock (Clepsydra) of Ktesibios," History of Physics, http://en.history-of-physics.com/antike/griechenland_wasseruhr.htm.

76 Your body also uses negative feedback: The technical term is *homeostasis*.

79 the Indian star tortoise, whose dome-shaped shell is shaped like the recently discovered *gömböc*: Adam Summers, "The Living Gömböc: Some Turtle Shells Evolved the Ideal Shape for Staying Upright," *Natural History* (March 2009), http://www.naturalhistorymag.com/biomechanics/10309/the-living-gomboc. Gömböcs are available for purchase at http://www.gomboc-shop.com/.

79 If you put a gömböc down on a horizontal surface: See the video at "Gomboc: How Turtles Self-Right," YouTube, January 28, 2008, http://www.youtube.com/watch?v=pn811yIALPw. The fascinating mathematical details are given in Gábor Domokos and Péter L. Várkonyi, "Geometry and Self-Righting of Turtles," *Proceedings of the Royal Society of London B* 275 (2008): 11–17.

81 Darrow lit a cigar into which he had inserted a wire: Louis B. Heller, *Do You Solemnly Swear? The Inside Story of How Cases Are Won and Lost* (New York: Doubleday & Co., 1968), quoted by E. E. Edgar, *Toledo Blade*, February 27, 1969, 45.

82 I was presenting a radio series: *How to Find the Sweet Spot* was broadcast on BBC Radio 4, September 6–10, 2004; see BBC Radio 4, "How to Find the Sweet Spot," http://www.bbc.co.uk/radio4/science/sweetspot.shtml; and "How to Find the Sweet Spot," Len Fisher: Science in Everyday Life, http://www.lenfisherscience.com/radio_and_tv/how_to_find_the_sweet_spot.html.

82 I had surfed on a tsunami: Admittedly it was a tiny tsunami—about three feet high! It was the "bore" that speeds up the Severn estuary in Somerset (United Kingdom) every month or so.

82 This extraordinary possibility was discovered by a scientist called A. Stephenson: A. Stephenson, "On a New Type of Dynamical Stability," *Memoirs and Proceedings of the Manchester Literary and Philosophical Society* 52 (1908): 1–10.

83 the conditions for driven equilibrium: Those conditions to keep a chain of linked rods balanced have been spelled out in detail by D. J. Acheson, "Multiple-Nodding Oscillations of a Driven Inverted Pendulum," *Proceedings of the Royal Society of London* 448 (1995): 89–95. See also D. J. Acheson, "A Pendulum Theorem," *Proceedings of the Royal Society of London* A443 (1993): 239–245.

83 the conditions . . . were realized in practice by . . . Tom Mullin: D. J. Acheson and Tom Mullin, "Upside-Down Pendulums," *Nature* 366 (1993): 215–216.

83 Tom later managed this, but with a wire rather than a rope: Tom Mullin et al., "The 'Indian Wire Trick' Via Parametric Excitation: A Comparison Between Theory and Experiment," *Proceedings of the Royal Society of London* A459 (2003): 539–546.

84 *control theory*, developed by . . . James Clerk Maxwell: James Clerk Maxwell, "On Governors," *Proceedings of the Royal Society of London* 16 (1868): 270–283.

85 the previously "impossible" device known as the Segway: It is possible to build your own Segway; see "The KIY Segway," http://web.mit.edu/first/segway/. This MIT website also has good details on how it works, but be aware that the commercial version has redundant safety mechanisms that are not available to the amateur.

85 the Segway is an *inverted pendulum*: For accessible scientific detail, see, for example, NationMaster.com, "Encyclopedia: Inverted Pendulum," http://www.statemaster.com/encyclopedia/Inverted-pendulum. For a practical, computer-controlled example, see "Inverted Pendulum Optimal Control," YouTube, June 16, 2008, http://www.youtube.com/watch?v=mqLI1d6R-Kc.

Modern computer software even allows us to balance two linked inverted pendulums on end. The theoretical conditions have been spelled out by Alexander Bogdanov, "Optimal Control of a Double Inverted Pendulum on a Cart," Oregon Health and Science

University technical report, December 2004, speech.bme.ogi.edu/publications/ps/bogdanov04a.pdf. For a practical example, see the video at http://www.youtube.com/watch?v=mqLI1d6R-Kc.

85 **although George W. Bush managed it:** See "Bush vs. Chimp: Segway Edition," YouTube, October 27, 2008, http://www.youtube.com/watch?v=hv-JcUyNgEs.

86 **[Adam Smith's] "invisible hand":** Adam Smith, *An Inquiry into the Nature and Causes of the Wealth of Nations* (1776, reprint, Chicago: University of Chicago Press, 1977). The full text is available as a Project Gutenberg e-book, http://www.gutenberg.org/files/3300/3300–8.txt.

Forbes ("The Lake as a Microcosm") seems to have borrowed from Smith's arguments when he says that "just as certainly as the thrifty business man who lives within his income will finally dispossess his shiftless competitor who can never pay his debts, the well-adjusted aquatic animal will in time crowd out its poorly-adjusted competitors for food and for the various goods of life."

Many authors have equated the "invisible hand" of democracy or capitalist economy with the evolution-driven "balance of nature." One of the earliest was George Perkins Marsh in *Man and Nature* (Cambridge, Mass.: Belknap Press of Harvard University Press, 1864).

86 **"Private Vice is Publick Benefit":** "Fable of the Bees" was originally published in 1701. Its prescient metaphorical description of democracy is discussed by David L. Norton in *Democracy and Moral Development: A Politics of Virtue* (Berkeley: University of California Press, 1995), 33.

87 **"after much trial and much error, we usually get it right":** Jon Meacham, "Democracy Is a Pesky Thing," *Newsweek*, March 15, 2010, http://www.newsweek.com/id/234582#CommentBox.

87 **Toynbee compared the evolution of societies to a musical rhythm:** Arnold Toynbee, *A Study of History* (Oxford: Oxford University Press, 1946), 549. Toynbee's actual description of the process was "rout-rally-rout-rally-rout-rally-rout." Harvard physics associate Stephen Blaha presents many examples in graphical form in "Reconstructing Prehistoric Civilizations in a New Theory of Civilizations," http://cogprints.org/2929/.

87 **Similar patterns are observed in banking systems:** One such pattern is the well-known "market cycle"; see, for example, Matt Blackman, "Market Cycles: The Key to Maximum Returns," Investopedia, http://www.investopedia.com/articles/technical/04/050504.asp.

90 **closer to the *punctuated equilibrium*:** Niles Eldredge and Stephen Jay Gould, "Punctuated Equilibria: An Alternative to Phyletic Gradualism," in *Models in Paleobiology*, edited by Thomas Schopf (San Francisco: Freeman, Cooper and Co., 1972), 82–115.

90 **G. K. Chesterton beat them to it:** G. K. Chesterton, "The Strange Crime of John Boulnois," in *The Father Brown Stories* (London: Book Club Associates, 1974), 292–304.

90 **Eldredge and Gould's real scientific theory went on to become a major strand of evolutionary thinking:** Stephen Jay Gould and Niles Eldredge, "Punctuated Equilibrium Comes of Age," *Nature* 366 (1993): 223–227.

CHAPTER 7: THE CHAOTIC ECOLOGY OF DRAGONS

95 **"Meddle not in the affairs of dragons":** This anonymous but widely quoted (on the Internet) line is a fairly obvious parody of Gildor the Elf's saying in J. R. R. Tolkien's *Lord of the Rings*, vol. 1, *The Fellowship of the Ring* (ch. 3): "Do not meddle in the affairs of wizards, for they are subtle and quick to anger."

95 **"The Ecology of Dragons" . . . has become a cult classic among ecologists:** Robert M. May, "The Ecology of Dragons," *Nature* 264 (1976): 16–17.

95 **a remarkable piece of research:** Peter Hogarth, "Ecological Aspects of Dragons," *Bulletin of the British Ecological Society* 7 (1976): 2–5. As well as evincing May's brilliant follow-up, Hogarth's seminal article also inspired some highly imaginative suggestions that appeared in a series of letters to the editor (see *Bulletin of the British Ecological Society* [December 1976]: 2–3). These included the suggestion that fire-breathing dragons may have become extinct because of fatal backfiring caused by "bursts of oxygen by catalase reactions—during periods of evolutionary trial and error—which were apt to cause disastrous internal explosions."

96 **the king had enacted legislation . . . forbidding any knight to . . . slay a local dragon:** These were the Knights Hospitaller—an order that has survived many vicissitudes throughout history to become today's medical and humanitarian Sovereign Military Order of Malta (http://www.orderofmalta.org.uk/).

96 **the supply of medieval knights . . . simply gave out:** Suggested by Holger Jannasch of the Woods Hole Oceanographic Institute in a letter to the editor of the *Bulletin of the British Ecological Society* (December 1976: 2).

96 **as pointed out by Holger Jannasch:** See *Bulletin of the British Ecological Society* (December 1976): 2. Another suggestion was that dragons simply lost credibility, as they did in Terry Pratchett's first "Discworld" novel, *The Color of Magic* (London: Corgi Books, 1985). They have probably been replaced in modern minds by other mysterious flying objects, such as UFOs.

96 **[May] was one of the first scientists to show how the mathematical methods used by physicists could . . . be applied to . . . ecology:** Robert May, *Stability and Complexity in Model Ecosystems* (Princeton, N.J.: Princeton University Press, 1973).

97 **The question [of population growth] had . . . been addressed . . . by the Rev. Thomas Malthus:** Thomas Malthus, *An Essay on the Principle of Population, as It Affects the Future Improvement of Society with Remarks on the Speculations of Mr. Godwin, M. Condorcet, and Other Writers* (London: J. Johnson, St. Paul's Church-Yard, 1798), available as a Project Gutenberg e-book, http://www.gutenberg.org/files/4239/4239.txt.

97 **[Malthus's] independent wealth and comfortable lifestyle:** Patricia James, *Population Malthus: His Life and Times* (New York: Routledge, 2006).

97 **His Cambridge degree:** Malthus's degree was a first-class degree, and he was listed as "Ninth Wrangler" in mathematics—a very respectable achievement.

97 **what might happen as the population of the world grew ever bigger:** "Population Growth over Human History," January 4, 2006, http://www.globalchange.umich.edu/globalchange2/current/lectures/human_pop/human_pop.html. The lecture outlined at this website gives a good account of the current state of play with regard to

estimates of the earth's long-term carrying capacity for the human species, which is generally placed at somewhere between 10 billion and 20 billion.

99 **The Belgian mathematician Pierre François Verhulst added a negative feedback term:** Pierre François Verhulst, "Notice sur la loi que la population pursuit dans son accroissement," *Correspondence Mathematique et Physique* 10 (1838): 113–121.

100 **"Th' whole worl's in a terrible state o' chassis":** When Captain Boyle utters his despairing cry at the end of Sean O'Casey's famous play about life in Dublin during the Irish Civil War, his whole world has collapsed in disorder and confusion. Nothing that he had predicted has worked out. His family has left him, a hoped-for inheritance has not eventuated, and he has just sixpence left in his pocket. Chaos truly seems to reign.

101 **apparently simple equation started to give crazy . . . results:** For a basic description, see my book *The Perfect Swarm*, 20–21. For a more extended description, see James Gleick, *Chaos* (London: William Heinemann, 1988), 69–76. There are many good iteration applets available to experiment with; see, for example, http://www.dallaway.com/pondlife/LogisticGraph.html.

102 **the oscillations began to go hopelessly out of control:** Gleick, *Chaos*, x.

102 **in a seminal article:** Robert M. May, "Simple Mathematical Models with Very Complicated Dynamics," *Nature* 261 (1976): 459–467, reproduced in full at http://nedwww.ipac.caltech.edu/level5/Sept01/May/May2.html. A wonderful read, even for a nonmathematician who can only read between the lines of the equations.

103 **natural ecosystems can have *many* possible stable states:** Robert M. May, "Thresholds and Breakpoints in Ecosystems with a Multiplicity of Stable States," *Nature* 269 (1977): 471–477.

103 **John Sutherland . . . confirm[ed] these predictions:** John P. Sutherland, "Multiple Stable Points in Natural Communities," *American Naturalist* 108 (1974): 859–873.

103 **Blackbeard's flagship *Queen Anne's Revenge:*** It ran aground on a sandbar in Beaufort Inlet (then known as Topsail Inlet) in 1718—oddly enough, Blackbeard was bringing her in to scrape the bottom clear of fouling organisms. The wreck is now the subject of a major conservation effort, the *Queen Anne's Revenge* Shipwreck Project (http://www.qaronline.org/).

104 **Imagine a pinball machine:** This description is adapted from that first produced by May, "Thresholds and Breakpoints in Ecosystems."

105 **flamingos are known not to breed unless their numbers and density are above a critical level:** Simon Pickering, Emma Crichton, and Barry Stevens-Wood, "Flock Size and Breeding Success in Flamingos," *Zoo Biology* 11 (1992): 229–234.

105 **the *Allee effect*:** Philip A. Stephens, William J. Sutherland, and R. P. Freckleton, "What Is the Allee Effect?" *Oikos* 87 (1999): 185–190, http://www.jstor.org/stable/3547011. See also Scheffer, *Critical Transitions in Nature and Society* (16–18, 196–199), for an excellent and simple discussion that includes some of the examples listed.

CHAPTER 8: TEETERING ON THE BRINK OF CATASTROPHE

109 **"Anything . . . changing steadily [is] heading toward catastrophe":** Susan Sontag, *AIDS and Its Metaphors* (New York: Penguin Books, 1990), ch. 8.

109 Catastrophe theory . . . [was] invented by the eccentric French mathematician René Thom: *Stabilité structurelle et morphogénèse* (New York: Benjamin, 1972); translated into English as *Structural Stability and Morphogenesis* (New York: Benjamin-Addison-Wesley, 1975). For a readable summary (which still requires a reasonable mathematical background), see Tim Poston and Ian Stewart, *Catastrophe Theory and Its Applications* (London: Pitman Publishing, 1978).

The origins of the word *catastrophe* lie in the classical Greek tragedies of playwrights such as Sophocles who showed the protagonists being drawn inexorably by circumstances and their own characters into disastrous situations. The culmination, or *catastrophe*, that ended the play involved the moral destruction or death of the central character.

109 [Einstein's theory of relativity] grabbed the public imagination: Not always favorably, as judged by the number of nonscientists who attempted to disprove it (Milena Wazeck, "Who Were Einstein's Opponents? Popular Opposition to the Theory of Relativity in the 1920s," Max Planck Institute for the History of Science, http://www.mpiwg-berlin.mpg.de/en/news/features/feature7).

110 D'Arcy Thompson's *On Growth and Form*: First published in 1917, *On Growth and Form* (Cambridge: Cambridge University Press) is now available in a full edition (some one thousand pages) online at: http://www.archive.org/details/ongrowthform1917thom.

110 I often could not make heads or tails of what I was reading: What was I to make, for example, of his mathematical description of how embryos develop?:

> Suppose that, in the model, the space U into which the growth wave is mapped, and which parametrizes the average biochemical state of each cell, is a four-dimensional state $\mathbf{R}^4$ identified with space-time. . . . if U actually has many more dimensions and we suppose that the growth wave $F(B^3, t)$ describing the evolution of the embryo, the only effective part of U (in embryogenesis) will be a four-dimensional domain.

The famous biologist Conrad Waddington writes in the foreword that this is where "the key point is . . . most clearly stated."

111 "My impression of René Thom": Francis Crick, *What Mad Pursuit: A Personal View of Scientific Discovery* (London: Penguin, 1988), 136.

111 Christopher Zeeman was on hand: E. Christopher Zeeman, "Catastrophe Theory," *Scientific American* 234 (1976): 65–83. Tim Poston and Ian Stewart subsequently produced an excellent guide, *Catastrophe Theory and Its Applications*, but it requires a degree of mathematical sophistication that places much (but not all) of its valuable material beyond the reach of nonmathematicians.

111 *all* catastrophes . . . can be classified into just seven "elementary" types: This is true for processes controlled by no more than four factors. Zeeman even built a "catastrophe machine"; see E. Christopher Zeeman, "A Catastrophe Machine," in *Towards a Theoretical Biology*, vol. 4, edited by Conrad H. Waddington (Edinburgh: Edinburgh University Press, 1972), 276–282. See also "Doctor Zeeman's Original Catastrophe Machine," http://www.math.sunysb.edu/~tony/whatsnew/column/catastrophe-0600/cusp4.html, and Daniel J. Cross, "Zeeman's Catastrophe Machine in Flash," http://lagrange.physics.drexel.edu/flash/zcm/. Poston and Stewart (*Catastrophe Theory and its Applications*) give detailed instructions on how to construct a catastrophe machine from rubber bands and cardboard.

111 except to mathematicians: Readers should be aware (as mathematicians certainly will be!) that I am simplifying Thom's theory in the interests of showing where its importance lies from the point of view of predicting disasters. As Poston and Stewart point out (*Catastrophe Theory and Its Applications*, 7), "Catastrophe theory is not a single thread of ideas; it resembles more closely a web with innumerable interconnected strands. . . . A proper perspective on the theory involves some appreciation of *all* of these strands and the way they combine. The elementary catastrophes . . . are but one strand, though an important one."

The elementary catastrophes, in other words, give the basic flavor of something much richer, which neither space nor expertise permits me to explore further. Fortunately, the basic flavor helps quite a lot when it comes to understanding the methods developed by scientists to forecast and anticipate disasters.

111 The cusp seems to be everywhere: As well as featuring in the reflection of light from a cylindrical surface, the cusp also can be detected in the scattering of light from ripples on the surface of water and even the "twinkling" of stars (Michael V. Berry, "Focusing and Twinkling: Critical Exponents from Catastrophes in Non-Gaussian Random Short Waves," *Journal of Physics A* 12 [1977]: 2061–2081).

112 One of this type's most inimical manifestations is the *poverty trap*: This is one of many real-life examples of the fold catastrophe offered by Martin Scheffer in *Critical Transitions in Nature and Society*.

114 the idea of *microcredit*: Sam Daley-Harris, "State of the Microcredit Summit Campaign" (Washington, D.C.: Microcredit Summit Campaign, 2009), https://promujer.org/espanol/dynamic/our_publications_5_Pdf_EN_SOCR2009%20English.pdf.

116 a clear shallow lake . . . suddenly becomes dark and turbid: Marten Scheffer, *Ecology of Shallow Lakes* (Berlin: Springer, 1997); Marten Scheffer and Egbert H. van Ness, "Shallow Lake Theory Revisited: Various Alternative Regimes Driven by Climate, Nutrients, Depth, and Lake Size," *Hydrobiologia* 584 (2007): 455–466.

116 Psychologists use it to help understand mood swings: Derek W. Scott, "Catastrophe Theory Applications in Clinical Psychology," *Current Psychology* 4 (1985): 69–86, http://www.springerlink.com/content/57631j8221752438/.

116 rigid patterns of behavior and even the collapse of ancient civilizations: Marten Scheffer and Frances R. Westley, "The Evolutionary Basis of Rigidity: Locks in Cells, Minds, and Society," *Ecology and Society* 12 (2007): 36, http://www.ecologyandsociety.org/vol12/iss2/art36/.

120 Konrad Lorenz's observations of the behavior of dogs: Konrad Lorenz, *On Aggression* (New York: Routledge, 2002).

120 our rather similar behavior as consumers: Terence A. Oliva and Alvin C. Burns, "Catastrophe Theory as a Model for Describing Consumer Behavior," *Advances in*

Consumer Research 5 (1978): 273–276, http://www.acrwebsite.org/volumes/display .asp?id=9434.

120 Earlier models made the naive assumption: These models are called the theory of reasoned action (Martin Fishbein and Icek Ajzen, *Beliefs, Attitudes, Intentions, and Behavior: An Introduction to Theory and Research* [Reading, Mass.: Addison-Wesley, 1975]) and the theory of planned behavior (Icek Ajzen, "From Intention to Actions: A Theory of Planned Behavior," in *Action Control: From Cognition to Behavior*, edited by Julius Kulh and Jurgen Beckman [Berlin: Springer-Verlag, 1985]: 11–39).

CHAPTER 9: MODELS AND SUPERMODELS

123 "Some say the world will end in fire": "Fire and Ice" was first published in *Harper's* 142, no. 847 (December 1920): 67. The critic John Serio asserts that Frost's poem is based on Dante's "Inferno"; he draws a parallel between the nine lines of the poem and the nine rings of Hell ("On 'Fire and Ice,'" Modern American Poetry, http://www .english.illinois.edu/maps/poets/a_f/frost/fireice.htm).

124 the German chemist August Kekulé: His full name was Friedrich August Kekulé von Stradonitz, but he apparently disliked his first name so much that he always called himself by the second.

124 "distinguished larger figures in manifold shapes": Kekulé gave this account during a speech to the German Chemical Society in 1890, twenty-five years after his discovery. Kekulé's written version of the speech has been translated into English several times, most recently by three different translators employed by the Harvard psychologist Albert Rothenberg. Rothenberg was interested in discovering whether Kekulé really was describing a dream; he came to the conclusion that Kekulé was more likely to have been in a "half-awake" state characterized by "homospatial" and "janusian" processes of creation (Albert Rothenberg, "Creative Cognitive Processes in Kekulé's Discovery of the Structure of the Benzene Molecule," *American Journal of Psychology* 108 [1995]: 419–438, http://www.jstor.org/stable/1422898).

125 cardboard cutouts . . . during the search for the DNA structure: James D. Watson, *The Double Helix: A Personal Account of the Discovery of the Structure of DNA* (London: Weidenfeld & Nicolson, 1968).

126 modern-day drug designers . . . use computer programs rather than cardboard: See, for example, Chun Meng Song, Shen Jean Lim, and Joo Chuan Tong, "Recent Advances in Computer-Aided Drug Design," *Briefings in Bioinformatics* 10 (2009): 579–591, http://bib.oxfordjournals.org/cgi/content/abstract/10/5/579; David C. Young, *Computational Drug Design: A Guide for Computational and Medicinal Chemists* (Hoboken, N.J.: John Wiley & Sons, 2009). One of the earliest successes was the antiglaucoma drug dorzolamide, introduced to the market by Merck in 1995 (http:// www.medicinenet.com/dorzolamide/article.htm). Another success was imatinib (http://www.macmillan.org.uk/Cancerinformation/Cancertreatment/Treatmenttypes/ Biologicaltherapies/Cancergrowthinhibitors/Imatinib.aspx), discovered by computer-based screening and trumpeted on the cover of *Time* magazine on May 28, 2001, as one of the "bullets" that were "new ammunition in the war against cancer" (http://www

.time.com/time/covers/0,16641,20010528,00.html). Imatinib has been very effective in treating the cancers for which it was designed, and there have been many other successes since for computer-aided drug design. Marketed as Gleevec, imatinib earned its discoverers a Lasker-DeBakey Clinical Medical Research Award in 2009 for "converting a fatal cancer into a chronic manageable condition" (Claudia Dreifus, "A Conversation with Brian J. Druker, M.D., Researcher Behind the Drug Gleevec," *New York Times*, November 2, 2009, http://www.nytimes.com/2009/11/03/science/03conv.html?_r=1).

126 continents . . . fit together like the pieces of a jigsaw puzzle: There are many good animations of the process. One that shows the "jigsaw puzzle" aspect well is at: http://www.structural-geology-portal.com/continental_drift_animation.html.

126 fruitful *predictions* that can be tested: Wegener's model suggested that there must be something carrying the continents around and led to the discovery of *tectonic plates*, while Kekulé's model predicted that all of the carbon atoms in the benzene molecule would be chemically equivalent. Both of these were important predictions, but Watson and Crick's model carried an extra bite because its predictions were *quantitative*. This meant that predictions based on it could be more detailed, and also that they could be more rigorously tested.

127 metaphors . . . for the structure of DNA: Sergi Cortiñas Rovira, "Metaphors of DNA: A Review of the Popularization Processes," *Journal of Science Communication* 7 (2008): 1–8.

127 the DNA of leadership: Judith E. Glaser, *The DNA of Leadership: Leverage Your Instincts to Communicate, Differentiate, Innovate* (New York: Adams Media Corp., 2006).

127 the DNA of innovation: See "The Innovator's DNA," INSEAD, December 21, 2009, http://knowledge.insead.edu/innovation-innovators-dna-091221.cfm?vid=358.

127 the DNA of customer experience: Colin Shaw, *The DNA of Customer Experience: How Emotions Drive Value* (New York: Palgrave-Macmillan, 2007).

127 a useful encapsulation: The most profitable encapsulation may have been that of Iranian designer Bijan Pakzad, who won a 1995 Ig Nobel Prize for creating DNA Cologne and DNA Perfume, neither of which contain DNA and both of which come in a *triple* helix bottle ("Winners of the Ig Nobel Prize," Improbable Research, http://improbable.com/ig/winners/#ig1995).

127 [Maslow's] hierarchy of human needs: Abraham H. Maslow, "A Theory of Human Motivation," *Psychological Review* 50 (1943): 370–396.

127 "one of psychology's genuinely good ideas": Christopher Peterson and Nansook Park, "What Happened to Self-Actualization?" *Perspectives on Psychological Science* 5 (2010): 320–322, http://pps.sagepub.com/content/5/3/320.full.

127 Maslow's model . . . has been criticized on grounds [of] ethnocentricity: Geert Hofstede, "The Cultural Relativity of the Quality of Life Concept," *Academy of Management Review* 9 (1984): 389–398, http://www.jstor.org/stable/258280; **psychological realism:** A. Wahba and L. Bridgewell, "Maslow Reconsidered: A Review of Research on the Need Hierarchy Theory," *Organizational Behavior and Human Performance* 15 (1976): 212–240; William G. Huitt, "Maslow's Hierarchy of Needs,"

Educational Psychology Interactive (2007), http://www.edpsycinteractive.org/topics/regsys/maslow.html; **position given to sex:** Douglas T. Kenrick, Vladas Griskevicius, Steven L. Neuberg, and Mark Schaller, "Renovating the Pyramid of Needs: Contemporary Extensions Built Upon Ancient Foundations," *Perspectives on Psychological Science* 5 (2010), http://www.csom.umn.edu/assets/144040.pdf.

128 Maslow . . . later revised his model: Mark E. Koltko-Rivera, "Rediscovering the Later Version of Maslow's Hierarchy of Needs: Self-Transcendence and Opportunities for Theory, Research, and Unification," *Review of General Psychology* 10 (2006): 302–317.

128 as happens with Maslow's model when it is invoked to "explain" the present concern of many people about the possibility of global warming: Philip Stott, "Global Warming: The Death of a Grand Narrative," Global Warming Policy Foundation, July 26, 2010, http://www.thegwpf.org/opinion-pros-a-cons/1305-global-warming-the-death-of-a-grand-narrative.html.

128 the metaphor of a greenhouse: After searching many websites for the best explanation, I find that most of them are incomplete or misleading. By far the best is the Wikipedia entry (http://en.wikipedia.org/wiki/Greenhouse_effect); it is one of the few that mentions the change in frequency/energy between visible light and infrared radiation that lies at the heart of the effect.

130 The models are mounted on a "shaker" table: The large-scale shaker table used by design engineers at the Pacific Earthquake Engineering Research Center in Berkeley, California, is twenty feet square and can support a sixty-ton model of a nine-story building ("Earthquake Simulator Laboratory," http://peer.berkeley.edu/laboratories/earthquake_simulator_lab.html). This table has been used to test a different solution: placing materials that can act as "roller bearings" beneath the foundations of a building, effectively decoupling the building from the lateral vibrations of the ground below (Farzad Naeim and James M. Kelly, *Design of Seismic Isolated Structures: From Theory to Practice* [New York: John Wiley & Sons, 1999]).

131 the . . . accuracy of weather forecasts is improving at roughly one day per decade: Personal communication (August 2010) from Dr. Marion Mittermaier, leader of the verification team responsible for monitoring the performance of the U.K. Met Office modeling system.

The formal measure of performance is the Numerical Weather Prediction (NWP) index, a multi-parameter measure—not susceptible to simple interpretation—that has improved from 100 to 118 over the past decade ("U.K. Computer Model Forecast Accuracy," Met Office, http://www.metoffice.gov.uk/corporate/verification/uk_nwp index.html). Unfortunately, it is not easy to compare the U.K. NWP index with the overall performance of weather forecasters in the United States because of the multiplicity of organizations and targets involved.

One big problem with weather forecasts is finding terms in which to express them that can be understood by the public. There is a debate as to whether frequency ("nine times out of ten") or probability ("90 percent reliable") is understood better by the layperson; at the moment the odds are on the latter (Susan L. Joslyn and Rebecca M. Nichols, "Probability or Frequency? Expressing Forecast Uncertainty in Public Weather Forecasts," *Meteorological Applications* 16 [2009]: 309–314).

For an interesting comparison with the simple rule that "tomorrow's weather will be the same as today's," see "How Good Are the Weather Forecasts?" http://weather.slimyhorror.com/.

131 rules . . . from game theory: Fisher, *Rock, Paper, Scissors.*

131 Bruce Bueno de Mesquita claims a 90 percent hit rate: Bruce Bueno de Mesquita, *The Predictioneer's Game : Using the Logic of Brazen Self-Interest to See and Shape the Future* (New York: Random House, 2009), http://www.predictioneersgame.com/.

132 people often act in a less rational, and sometimes more altruistic, manner: Fisher, *Rock, Paper, Scissors*, ch. 5.

132 a "safe operating space for humanity": Johan Rockström et al., "A Safe Operating Space for Humanity," *Nature* 461 (2009): 472–475. You can follow the debate on this issue, and also join it, at Climate Feedback, "Planetary Boundaries," http://blogs.nature.com/climatefeedback/2009/09/planetary_boundaries.html. The full scientific report can be accessed at Stockholm Resilience Centre, "Tipping Towards the Unknown," http://www.stockholmresilience.org/planetary-boundaries.

134 *Gaia hypothesis* (which is itself a model): The Gaia hypothesis was first advanced by James Lovelock, a highly respected NASA scientist, in the mid-1960s (James E. Lovelock, "A Physical Basis for Life Detection Experiments," *Nature* 207 [1965]: 568–570). It came to public attention with the publication of Lovelock's popular book *Gaia: A New Look at Life on Earth* (Oxford: Oxford University Press, 1979) and has been the center of controversy ever since. One of the most trenchant critics has been the Berkeley scientist James Kirchner (see, for example, James W. Kirchner, "The Gaia Hypothesis: Fact, Theory, and Wishful Thinking," *Climatic Change* 52 [2002]: 391–408). Kirchner's main claim is that "in the real world . . . natural selection favors any trait that gives its carriers a reproductive advantage over its non-carriers, whether it improves or degrades the environment (and thereby benefits or hinders its carriers and non-carriers alike). Thus Gaian and anti-Gaian feedbacks are both likely to evolve."

134 a single gigantic organism with self-regulating processes: Timothy M. Lenton and Marcel van Oijen, "Gaia as a Complex Adaptive System," *Philosophical Transactions of the Royal Society of London B* 357 (2002): 683–695.

CHAPTER 10: BEWARE OF MATHEMATICIANS

135 "Beware of mathematicians": Saint Augustine, *De genesi ad litteram*, vol. 2, ch. 17, translation published in Morris Kline, *Mathematics in Western Culture* (New York: Oxford University Press, 1953), 3; later retranslated by J. H. Taylor in *Ancient Christian Writers 41: Saint Augustine* (London: Newman Press, 1982) as: "Avoid astrologers and all impious soothsayers, especially when they tell the truth."

Probably the second translation is the more accurate (and almost directly opposite in meaning to the first!), but I don't really care—both versions are extraordinarily apposite to this book.

136 "garbage in, garbage out" [is] a phrase . . . usually credited to . . . George Fuechsel: See WiseGeek, "What Is Garbage In, Garbage Out?," http://www.wisegeek.com/what-is-garbage-in-garbage-out.htm. I was amused to find while researching this section that GIGO is also a logo for a brand of men's underwear (http://www.gigo.com.co/).

136 **the unexpectedly short maiden flight of . . . Ariane 5:** Peter B. Ladkin, "The Ariane 5 Accident: A Programming Problem?" Bielefeld University article RVS-J-98-02, May 24, 2002, http://www.rvs.uni-bielefeld.de/publications/Reports/ariane.html.

137 **When the Newfoundland cod fisheries closed in 1992:** See Stephen R. Carpenter, Carl Folke, Marten Scheffer, and Frances Westley, "Resilience: Accounting for the Non-computable," *Ecology and Society* 14, no. 1 (2008): 13.

137 **reports . . . [from] war zones . . . are frequently over-optimistic:** Richard K. Betts, "Analysis, War, and Decision: Why Intelligence Failures Are Inevitable," *World Politics* 31 (1978): 61–79, http://www.jstor.org/stable/2009967; see the example on p. 68. This article is a highly recommended read and provides good evidence that the phrase "military intelligence" is indeed an oxymoron.

137 **[businesses] provide the modeler with selective information:** Donella H. Meadows and Jennifer M. Robinson, "The Electronic Oracle: Computer Models and Social Decisions," *System Dynamics Review* 18 (2002): 271–308.

137 **claimed that some law schools deliberately inflate predicted graduate incomes:** "GIGO Means Garbage In, Garbage Out," Lawyers Against the Law School Scam, July 26, 2010, http://lawschoolscam.blogspot.com/2010/07/gigo-means-garbage-in-garbage-out.html.

137 **Nolan McCarty . . . attributed the success . . . to his careful selection of input data:** Sanjida O'Connell, "The Predictioneer: Using Games to See the Future," *New Scientist* 2752 (March 17, 2010): 42–45.

138 **fewer data can lead to *more* accurate predictions:** See, for example, Gerd Gigerenzer and Henry Brighton, "Homo Heuristicus: Why Biased Minds Make Better Inferences," *Topics in Cognitive Science* 1 (2009): 107–143; Gerd Gigerenzer and Daniel Goldstein, "Models of Ecological Rationality: The Recognition Heuristic," *Psychological Review* 109 (2002): 75–90; Gerd Gigerenzer, Peter M. Todd, and ABC Research Group, *Simple Heuristics That Make Us Smart* (Oxford: Oxford University Press, 1999), 43.

138 **ecologists . . . [use] a process called "tuning":** Marten Scheffer and Jeroen Beets, "Ecological Models and the Pitfalls of Causality," *Hydrobiologia* 275–276 (1994): 115–124.

139 **minimalist models for decision-making:** Gigerenzer, Todd, and ABC Research Group, *Simple Heuristics That Make Us Smart.*

140 **developed by three MIT scientists and publicized in the 1972 book:** Donella H. Meadows, Dennis L. Meadows, Jørgen Randers, and William W. Behrens III, *The Limits to Growth* (New York: Universe Books, 1972). I have to declare an interest here: I was enthused by the book at the time of publication and gave many talks on its predictions. I often found myself on a collision course with other speakers, who seemed either not to have read the book or to have misinterpreted what they had read. Some even seemed to believe that it predicted the collapse of the entire world system by the end of the twentieth century—something that obviously has not happened and that the book never predicted *would* happen (Roger-Maurice Bonnet and Lodewyk Woltjer, *Surviving One Thousand Centuries: Can We Do It?* Berlin: Springer-Praxis, 2008).

140 **"only in the most limited sense of the word":** Meadows et al., *The Limits to Growth.*

141 "Analysis shows that 30 years of historical data": Graham M. Turner, "A Comparison of *The Limits to Growth* with Thirty Years of Reality," *Global Environmental Change* 18 (2008): 397–411, www.csiro.au/files/files/plje.pdf.

141 Many later models . . . have made similar predictions: Robert Costanza, Rik Leemans, Roelof Boumans, and Erica Gaddis, "Integrated Global Models," in *Sustainability or Collapse: An Integrated History and Future of People on Earth*, edited by Robert Costanza, Lisa J. Graumlich, and Will Steffen (Cambridge, Mass.: MIT Press, 2007), 417–446, http://74.125.155.132/scholar?q=cache:HtMRavIBN90J:scholar.google.com/&hl=en&as_sdt=2000.

141 Martin Gardner . . . claimed to have *dis*proved the theorem: For details of the four-color theorem and Gardner's supposed "disproof," see Wolfram MathWorld, "Four-Color Theorem," http://mathworld.wolfram.com/Four-ColorTheorem.html.

141 Kenneth Appel and Wolfgang Haken . . . proof: For a popular summary by these authors, see "The Solution of the Four-Color Map Problem," *Scientific American* 237 (1977): 108–121.

142 other proofs have emerged: See, for example, Neil Robertson, Daniel P. Sanders, Paul Seymour, and Robin Thomas, "The Four Color Theorem," November 13, 1995, http://people.math.gatech.edu/~thomas/FC/fourcolor.html#Outline.

142 Could there be a sociological equivalent of Newton's laws: Here is my personal version of Newton's three laws, applied to society:

1. *The Law of Inertia:* Things keep on bobbing along as they are unless some dramatic external event forces a change.
2. *The Law of Momentum:* Strong governance induces proportional change in people's behavior. (This version was suggested by Dr. Graham Turner.)
3. *The Law of Reaction:* For every social action there is an equal and opposite reaction.

142 *psychohistory* [is] "that branch of mathematics": Isaac Asimov, *Foundation* (London: Panther Books, 1960), 16.

143 the well-known "Prisoner's Dilemma": William Poundstone, *Prisoner's Dilemma* (Oxford: Oxford University Press, 1993).

143 social *thermodynamics*—the quest for social parallels to physical quantities: Josip Stepanic Jr. et al., "Approach to a Quantitative Description of Social Systems Based on Thermodynamic Formalism," *Entropy* 2 (2000): 98–105, http://74.125.155.132/scholar?q=cache:8h3ZBwqBr7IJ:scholar.google.com/&hl=en&as_sdt=2000.

143 the social organization of his home country was essentially perfect: Yi-Fang Chang, "Social Synergetics, Social Physics, and Research of Fundamental Laws in Social Complex Systems," Arxiv preprint arXiv:0911.1155 (2009), http://arxiv.org/pdf/0911.1155.

143 These efforts . . . sometimes attempt to use insights from neuroscience and evolutionary psychology: Bruce MacLennan, "Evolutionary Psychology, Complex Systems, and Social Theory," http://www.cs.utk.edu/~mclennan/papers/EPCSST.pdf.

143 United Nations Intergovernmental Panel on Climate Change: The ongoing assessment reports are available at the IPCC website, http://www.ipcc.ch/publications_

and_data/publications_and_data_reports.htm#2. For the most recent update, prepared by a prestigious group of national scientific academies, see Climate-L.org, "InterAcademy Council Delivers IPCC Review Report," August 30, 2010, http://climate-l.org/2010/08/31/interacademy-council-delivers-ipcc-review-report/. This report, prepared in the wake of published errors about the rate of melting of Himalayan glaciers, confirms that the IPCC is "at least 90% certain" that mankind is driving global warming (ABC News, "UN Hopes Science Review Eases Climate Skepticism," August 30, 2010, http://www.abc.net.au/news/stories/2010/08/29/2996648.htm).

144 a high-level meeting: The final report of this important meeting is by John Kambhu, Scott Weidman, and Neel Krishnan, "New Directions for Understanding Systemic Risk: A Report on a Conference Cosponsored by the Federal Reserve Bank of New York and the National Academy of Sciences," *Economic Policy Review* (2007): i–83, http://econpapers.repec.org/article/fipfednep/y_3a2007_3ai_3anov_3ap_3ai-83_3an_3av.13no.2.htm.

144 The outcome was summed up in an article: Robert M. May, Simon Levin, and George Sugihara, "Ecology for Bankers," *Nature* 451 (2008): 893–895.

145 banks of the future should not be staffed with ex-rocket scientists . . . but with ecologists: Sumit Paul-Choudhury, "How 'Rocket Science' Failed the Banks," *New Scientist* 200, no. 2687 (December 17, 2008): 38.

145 This is not to say that the individual models are perfect: Max Rietkerk et al., "Local Ecosystem Feedbacks and Critical Transitions in the Climate," *Biogeosciences Discussions* 6 (2009): 10121–10136.

145 the sort of economic modeling criticized by . . . Taleb and . . . Makridakis: Makridakis and Taleb, "Decision Making and Planning Under Low Levels of Predictability."

146 as one group of prominent ecologists has pointed out: Carpenter et al., "Resilience: Accounting for the Noncomputable." Much of the discussion in this section is based on this seminal article.

147 the tendency of scientists to focus on the things that they can measure and calculate and to ignore the rest: It doesn't just happen with modelers. Once you catch on to the basic idea, you can see it on the TV news any night of the week, coming out of the mouths of protesters, politicians, policymakers, and anyone who has an ax to grind. I call it the "black box" approach: Everything inside the box is given a value, and everything outside the box has zero value. What is actually inside the box depends on the person. It may be an issue, a cost, or, in the case of modelers, a set of physical parameters. The important point is that the stuff inside the box counts, and the stuff outside the box doesn't.

147 The story of Robert Millikan's Nobel Prize–winning measurement: For details, see my book *Weighing the Soul*, p. 10.

147 "The most exciting phrase to hear in science": Attributed to Isaac Asimov (source unknown).

147 studies of swarm intelligence: See my book *The Perfect Swarm.*

148 The subsequent investigation of the Himalayan glaciers controversy: Jeff Tollefson, "Climate Panel Must Adapt to Survive," *Nature* 467 (2010): 14.

148 "To err is human. To really foul up—it takes a computer": According to Fred Shapiro, editor of *The Yale Book of Quotations*, this well-known saying first appeared in the *Newark Advocate* in 1969 (Stephen J. Dubner, "Our Daily Bleg: Where Did 'Garbage' and 'Bugs' Come From?" Freakonomics, May 1, 2008, http://freakonomics.blogs.nytimes.com/2008/05/01/our-daily-bleg-where-did-garbage-and-bugs-come-from/).

148 According to one U.S. federal judge: Stanley A. Kurzban, "Authentication of Computer-Generated Evidence in the United States Federal Courts," *Idea* 237 (1994–1995): 17, http://heinonline.org/HOL/LandingPage?collection=journals&handle=hein.journals/idea35&div=30&id=&page=.

CHAPTER 11: WEAK SIGNALS AS MAJOR EARLY-WARNING SIGNS

151 "Every calamity is a spur and valuable hint": Ralph Waldo Emerson, *The Conduct of Life* (1860), http://infomotions.com/alex2/authors/emerson-ralph/emerson-conduct-752/.

151 snatch of dialogue: Reconstructed from several real-life examples of the type that most readers are likely to be familiar with.

152 Psychologists have identified weak signals in the form of stressful life events: Thomas H. Holmes and Richard H. Rahe, "The Social Readjustment Rating Scale," *Journal of Psychosomatic Research* 11 (1967): 213–218; Christopher Tennant and Gavin Andrews, "A Scale to Measure the Stress of Life Events," *Australian and New Zealand Journal of Psychiatry* 10 (1976): 27–32, http://informahealthcare.com/doi/abs/10.3109/00048677609159482.

152 there is general agreement: Daniel Weiss, "The Impact of the Event Scale: Revised," *International and Cultural Psychology*, part 2 (2007): 219–238, http://www.springerlink.com/content/vm53904v674j3283/; Scott Monroe, "Modern Approaches to Conceptualizing and Measuring Human Life Stress," *Annual Reviews of Clinical Psychology* 4 (2008): 33–52, http://www.annualreviews.org/doi/abs/10.1146/annurev.clinpsy.4.022007.141207.

152 Ansoff maintained that managers need a "weak signal mentality": Igor Ansoff, "Managing Strategic Surprise by Response to Weak Signals," *California Management Review* 18 (1975): 21–33.

152 events that business strategists label as "wild cards": Sandro Mendonça, Miguel Pina e Cunha, Jari Kaivo-oja, and Frank Ruff, "Wild Cards, Weak Signals, and Organizational Improvement," *Futures* 36 (2004): 201–218, http://fesrvsd.fe.unl.pt/WPFEUNL/WP2003/wp432.pdf.

152 Taleb's "black swans": Taleb, *The Black Swan*.

153 the explosion of NASA's *Challenger* space shuttle in 1986: For the full story, see Fisher, *The Perfect Swarm*, 95–98.

153 "a crisis sends off a repeated and persistent trail of early warning signals": Ian I. Mitroff, "Crisis Management: Cutting Through the Confusion," *Sloan Management Review* (Winter 1988): 18.

154 build a capacity into the organization *in advance* for improvisation: Mendonça et al., "Wild Cards, Weak Signals, and Organizational Improvement." Hierarchical organizational cultures can make this difficult, particularly if a hierarchical structure is also part of the national culture (Rohit Deshpandé, John U. Farley, and Frederick E. Webster, "Corporate Culture, Customer Orientation, and Innovativeness in Japanese Firms: A Quadrad Analysis," *Journal of Marketing* 57 [1993]: 23–37).

154 it is nearly impossible to invent crisis management mechanisms while the crisis is taking place: Mitroff, "Crisis Management."

155 Casti . . . claims that . . . "social mood" can be used as [a predictor]: John Casti, *Mood Matters: From Rising Skirt Lengths to the Collapse of World Powers* (New York: Springer, 2010).

155 The hemline index was proposed . . . by . . . George Taylor: See Paul H. Nystrom, *Economics of Fashion* (New York: Ronald Press, 1928). Many other social indices have been suggested, including the lipstick index (according to the chairman of Estée Lauder, analyzing sales data after the 9/11 attacks, lipstick sales *increase* in bad times because people want to feel good about themselves) and the men's underwear index (according to Alan Greenspan, former chair of the U.S. Federal Reserve and author of *The Age of Turbulence*, a dip in the sale of men's underwear is a sign that people are feeling the pinch) (Lala Rimando, "Economic Forecasting for Dummies," ABS-CBNnews.com, April 1, 2010, http://www.abs-cbnnews.com/business/01/04/10/economic-forecasting-dummies-check-hemlines-men%E2%80%99s-underwear).

155 "it is how a group or population sees the future that shapes events": John Casti, "How Social Mood Moves the World," *New Scientist* (May 22, 2010): 30–31.

155 skirt lengths *follow* changes in the economy: Marjolein van Baardwijk and Philip Hans Franses, "The Hemline and the Economy: Is There Any Match?" Erasmus School of Economics Econometric Institute Report 2010–40 (2010), http://publishing.eur.nl/ir/repub/asset/20147/EI%202010–40.pdf.

155 a sense of "environmental turbulence": Mendonça et al., "Wild Cards, Weak Signals, and Organizational Improvement," 211.

155 Ecologists have similarly identified "environmental turbulence": Marten Scheffer et al., "Early-Warning Signals for Critical Transitions," *Nature* 461 (2009): 53–59; William A. Brock, Stephen R. Carpenter, and Marten Scheffer, "Regime Shifts, Environmental Signals, Uncertainty, and Policy Choice," in *Complexity Theory for a Sustainable Future*, edited by Jon Norberg and Graeme Cumming (New York: Columbia University Press, 2010), 180–206.

These signs are similar in principle to those that were first observed by physicists studying transitions between different states in materials—transitions that could be as simple as melting or boiling or as complicated as the behavior of a material near its "triple point": the point where it can exist simultaneously as gas, liquid, and solid in a fluctuating, fuzzy equilibrium.

In these cases, it is the mutual interactions of atoms and molecules that produce the warning signs. In complex societies, economies, and ecosystems, the interactions between individual components are also the guiding factor, but some of these components are now living organisms, which greatly adds to the complexity of the problem.

156 an "eco-friendly" dress: Anders Mellbratt and Nils Wiberg, "The Fallacy of Eco-Friendly Design: How Progressive Design Can Save It," http://www.iasdr2009.org/ap/Papers/Orally%20Presented%20Papers/Sustainability/The%20Fallacy%20of%20Eco-Friendly%20Design%20-%20And%20how%20progressive%20design%20can%20save%20it.pdf.

156 Fold catastrophes have been shown to mimic . . . real-life critical transitions: Inappropriate use of this model and overclaiming of results caused some controversy in the early days of catastrophe theory (Raphael S. Zahler and Hector J. Sussman, "Claims and Accomplishments of Applied Catastrophe Theory," *Nature* 269 [1977]: 759–763). These issues are now resolved, and the value of the model is increasingly recognized (Scheffer et al., "Early-Warning Signals for Critical Transitions").

156 the key factor is *loss of resilience*: Carl Folke, "Resilience: The Emergence of a Perspective for Social-Ecological Change," *Global Environmental Change* 16 (2006): 253–267; Carl Folke et al., "Regime Shifts, Resilience, and Biodiversity in Ecosystem Management," *Annual Review of Ecology and Systematics* 35 (2004): 557–581; Brian Walker et al., "Resilience, Adaptability, and Transformability in Social-Ecological Systems," *Ecology and Society* 9 (2004); Brian Walker et al., "A Handful of Heuristics and Some Propositions for Understanding Resilience in Social-Economic Systems," *Ecology and Society* 11 (2006).

156 resilience [is] the capacity to absorb shocks and still function: Crawford S. Holling, "Resilience and Stability of Ecological Systems," *Annual Review of Ecology and Systematics* 4 (1973): 1–23.

156 "the capacity of a system to absorb disturbance": Walker et al., "A Handful of Heuristics and Some Propositions for Understanding Resilience in Social-Economic Systems." There is also a distinction to be drawn between resilience due to the width of the basin of attraction (called *ecological resilience*) and resilience as measured by the speed of return to the original state after a disturbance (*engineering resilience*) (Egbert H. van Nes and Marten Scheffer, "Slow Recovery from Perturbations as a Generic Indicator of a Nearby Catastrophic Shift," *The American Naturalist* 169 [2007]: 738–747). Although it can sometimes be important, I have glossed over this distinction in the interests of simplicity.

157 Mathematicians have identified five key early-warning signs: Scheffer et al., "Early-Warning Signals for Critical Transitions."

157 The evidence for their significance has largely emerged: Marten Scheffer and Stephen R. Carpenter, "Catastrophic Regime Shifts in Ecosystems: Linking Theory to Observation," *TRENDS in Ecology and Evolution* 18 (2003): 648–656, http://www.imedea.uib-csic.es/master/cambioglobal/Modulo_3_03/Scheffer%20and%20Carpenter%20regime%20shifts.pdf. A key, and very readable, article.

157 The evidence for their significance . . . is equally applicable: Scheffer et al., "Early-Warning Signals for Critical Transitions."

157 Another example is coral reefs: Magnus Nystrom et al., "Coral Reef Disturbance and Resilience in a Human-Dominated Environment," *Trends in Ecology and Evolution* 15 (2000): 413–417.

158 **The concept was introduced to ecology by . . . Holling:** Holling, "Resilience and Stability of Ecological Systems."

158 **We may represent [homeothermy] . . . by a one-dimensional "landscape":** Adapted from a suggestion in Walker et al., "Resilience, Adaptability, and Transformability in Social-Ecological Systems."

158 **a loss of resilience may arise from cultural conservatism:** Walker et al., "Resilience, Adaptability and Transformability in Social-Ecological Systems."

160 **One of the more curious factors:** See, for example, John Platt, "South African Gamblers Smoke Endangered Vulture Brains for Luck," in "Extinction Countdown," *Scientific American*, June 10, 2010, http://www.scientificamerican.com/blog/post.cfm?id=south-african-gamblers-smoke-endang-2010–06–10.

160 **These factors have been identified in a series of case studies:** Brian Walker and Rochelle L. Lawson, "E&S Special Feature: Case Studies in Resilience: Fifteen Social-Ecological Systems Across Continents and Societies," Resilience Alliance, http://www.resalliance.org/1613.php.

160 **One example . . . occurs in the Florida Everglades:** Lance Gunderson, "Everglades, Florida, USA," in Walker and Lawson, "E&S Special Feature: Case Studies in Resilience."

160 **"wild cards" for environmental change:** Stephen Carpenter and William Brock, "Adaptive Capacity and Traps," *Ecology and Society* 13 (2008): 40, http://www.ecologyandsociety.org/vol13/iss2/art40/.

161 **studies on model fisheries have shown that such action could indeed help us to turn back from the brink:** Reinette Biggs, Stephen R. Carpenter, and William A. Brock, "Turning Back from the Brink: Detecting an Impending Regime Shift in Time to Avert It," *Proceedings of the National Academy of Sciences* 106 (2009): 826–831, http://www.pnas.org/content/106/3/826.abstract.

162 **the place where flickering was shown to be an unambiguous warning sign:** Judi E. Hewitt and Simon F. Thrush, "Empirical Evidence of an Approaching Alternate State Produced by Intrinsic Community Dynamics, Climatic Variability, and Management Actions," *Marine Ecology Progress Series* 413 (2010): 267–276, http://www.int-res.com/articles/theme/m413_ThemeSection.pdf#page=84.

163 **the on-again-off-again relationship of celebrity couple Marilyn Manson and Evan Rachel Wood:** Joanna Sloame, "Favorite On-Again-Off-Again Celebrity Couples," *New York Daily News*, http://www.nydailynews.com/gossip/galleries/favorite_onandoff_celebrity_couples/favorite_onandoff_celebrity_couples.html.

163 **lakes are likely to suddenly "flip":** Marten Scheffer and Erik Jeppesen, "Regime Shifts in Shallow Lakes," *Ecosystems* 10 (2007): 1–3, http://www.springerlink.com/content/p81802346gh15771/.

163 **phosphorus levels in the water begin to fluctuate more and more widely:** Stephen Carpenter and William "Buz" Brock, "Rising Variance: A Leading Indicator of Ecological Transition," *Ecology Letters* 9 (2006): 311–318.

163 **global warming will increase the frequency of such extreme blocking patterns:** Christophe Cassou and Éric Guilyardi, "Modes de variabilité et changement climatique," in *La Météorologie* (2007), http://www.insu.cnrs.fr/f1745pdf,modes-variabilite-changement-climatique.pdf.

164 critical slowing down . . . is a characteristic of reduced resilience: Van Nes and Scheffer, "Slow Recovery from Perturbations as a Generic Indicator of a Nearby Catastrophic Shift."

164 The British humorist C. Northcote Parkinson: Parkinson's Law also states that "work expands so as to fill the time available for its completion." This "law" first appeared in an article in *The Economist* (November 19, 1955), which the author later expanded into a book entitled *Parkinson's Law or the Pursuit of Progress* (London: John Murray, 1958).

165 [critical slowing down] has been mathematically demonstrated to be an early-warning sign: Van Nes and Scheffer, "Slow Recovery from Perturbations as a Generic Indicator of a Nearby Catastrophic Shift."

165 self-organized banding in beds of seagrass: Tjisse van der Heide et al., "Spatial Self-Organized Patterning in Beds of Seagrasses Along a Depth Gradient of an Intertidal Ecosystem," *Ecology* 91 (2010): 362–369.

166 changes in the spatial patterns can be a signal of upcoming critical transitions: Vasilis Dakos et al., "Spatial Correlation as Leading Indicator of Catastrophic Shifts," *Theoretical Ecology* 3 (2009): 163–174.

166 the pattern of vegetation slowly becomes patchier in a very predictable way: Max Rietkerk, Stefan C. Dekker, Peter C. de Ruiter, and Johan van Koppel, "Self-Organized Patchiness and Catastrophic Shifts in Ecosystems," *Science* 305 (2004): 1926–1929, http://www.to.isac.cnr.it/aosta_old/aosta2005/LecturesSeminars/Rietkerk_I.pdf. Don't miss the wonderful images in this article.

167 yet one more [hint of upcoming catastrophe]—increasing skewness: Vishwesha Guttal and Ciriyam Jayaprakash, "Changing Skewness: An Early Warning Signal of Regime Shifts in Ecosystems," *Ecology Letters* 11 (2008): 450–460.

167 "Prediction is very difficult": Attributed to Niels Bohr in Arthur K. Ellis, *Teaching and Learning Elementary Social Studies* (London: Allyn & Bacon, 1970), 431.

167 at least a 90 percent chance that humankind's activities are making a substantial contribution: Climate-L.org, "InterAcademy Council Delivers PICC Review Report."

168 In the face of these odds: We should also be aware that some regime shifts in nature can occur *without* warning (Alan Hastings and Derin B. Wysham, "Regime Shifts in Ecological Systems Can Occur with No Warning," *Ecology Letters* 13 [2010]: 464–472).

INDEX

4/11